Reconnaissance of Contaminants in Selected Wastewater-Treatment-Plant Effluent and Stormwater Runoff Entering the Columbia River, Columbia River Basin, Washington and Oregon, 2008–10

By Jennifer L. Morace

Prepared in cooperation with the Columbia River Inter-Tribal Fish Commission and the Lower Columbia Estuary Partnership

Scientific Investigations Report 2012–5068

U.S. Department of the Interior
U.S. Geological Survey

U.S. Department of the Interior
KEN SALAZAR, Secretary

U.S. Geological Survey
Marcia K. McNutt, Director

U.S. Geological Survey, Reston, Virginia: 2012

For more information on the USGS—the Federal source for science about the Earth, its natural and living resources, natural hazards, and the environment, visit http://www.usgs.gov or call 1–888–ASK–USGS.

For an overview of USGS information products, including maps, imagery, and publications, visit http://www.usgs.gov/pubprod

To order this and other USGS information products, visit http://store.usgs.gov

Suggested citation:
Morace, J.L., 2012, Reconnaissance of contaminants in selected wastewater-treatment-plant effluent and stormwater runoff entering the Columbia River, Columbia River Basin, Washington and Oregon, 2008–10: U.S. Geological Survey Scientific Investigations Report 2012–5068, 68 p.

Contents

Figures

Tables

Tables—Continued

Conversion Factors, Datums, and Abbreviations and Acronyms

Conversion Factors

Inch/Pound to SI

Multiply	By	To obtain
foot (ft)	0.3048	meter (m)
mile (mi)	1.609	kilometer (km)
square mile (mi^2)	2.590	square kilometer (km^2)
cubic foot per second (ft^3/s)	0.02832	cubic meter per second (m^3/s)
million gallons per day (Mgal/d)	1.547	cubic meter per second (m^3/s)

SI to Inch/Pound

Multiply	By	To obtain
gram per day (g/d)	0.03527	ounce, avoirdupois (oz)
liter (L)	0.2642	gallon (gal)
liter per day (L/d)	0.2642	gallon per day (gal/d)

Temperature in degrees Celsius (°C) may be converted to degrees Fahrenheit (°F) as follows:

$$°F=(1.8×°C)+32.$$

Concentrations of chemical constituents in water are given either in milligrams per liter (mg/L), micrograms per liter (µg/L), or nanograms per liter (ng/L).

Specific conductance is given in microsiemens per centimeter at 25 degrees Celsius (µS/cm at 25 °C).

CAS Registry Numbers® is a Registered Trademark of the American Chemical Society. CAS recommends the verification of the CASRNs through CAS Client Services[SM].

Datums

Vertical coordinate information is referenced to the North American Vertical Datum of 1988 (NAVD 88).

Horizontal coordinate information is referenced to the North American Datum of 1983 (NAD 83).

Conversion Factors, Datums, and Abbreviations and Acronyms—Continued

Abbreviations and Acronyms

7Q10	the lowest streamflow for seven consecutive days that occurs on average once every 10 years
AOC	anthropogenic organic compounds
CAS	Chemical Abstracts Service
CERC	Columbia Environmental Research Center
CSO	Combined sewer overflow
E	estimated
EEQ	estradiol equivalent factor
EPA	U.S. Environmental Protection Agency
GC/MS	gas chromatography/mass spectrometry
MDL	Method detection limit
NPDES	National Pollutant Discharge Elimination System
NWQL	National Water-Quality Laboratory
ODEQ	Oregon Department of Environmental Quality
PAH	Polycyclic aromatic hydrocarbons
PBDE	Polybrominated diphenyl ethers
PCB	Polychlorinated biphenyls
PIL	plan initiation level (SB 737)
QC	quality control
RL	Reporting limit
RPD	relative percent difference
SB	Senate Bill
USGS	U.S. Geological Survey
WWTP	Wastewater-treatment plant
YES	Yeast estrogen screen

Reconnaissance of Contaminants in Selected Wastewater-Treatment-Plant Effluent and Stormwater Runoff Entering the Columbia River, Columbia River Basin, Washington and Oregon, 2008–10

By Jennifer L. Morace

Abstract

Toxic contamination is a significant concern in the Columbia River Basin in Washington and Oregon. To help water managers and policy makers in decision making about future sampling efforts and toxic-reduction activities, a reconnaissance was done to assess contaminant concentrations directly contributed to the Columbia River through wastewater-treatment-plant (WWTP) effluent and stormwater runoff from adjacent urban environments and to evaluate instantaneous loadings to the Columbia River Basin from these inputs.

Nine cities were selected in Oregon and Washington to provide diversity in physical setting, climate characteristics, and population density—Wenatchee, Richland, Umatilla, The Dalles, Hood River, Portland, Vancouver, St. Helens, and Longview. Samples were collected from a WWTP in each city and analyzed for anthropogenic organic compounds, pharmaceuticals, polychlorinated biphenyls (PCBs), polybrominated diphenyl ether (PBDEs [brominated flame-retardants]), organochlorine or legacy compounds, currently used pesticides, mercury, and estrogenicity. Of the 210 compounds analyzed in the WWTP-effluent samples, 112 (53 percent) were detected, and the detection rate for most compound classes was greater than 80 percent. Despite the differences in location, population, treatment type, and plant size, detection frequencies were similar for many of the compounds detected among the WWTPs. By contrast, the occurrence of polycyclic aromatic hydrocarbons (PAHs) was sporadic, and PCBs were detected at only three WWTPs.

The stormwater-runoff samples were analyzed for a slightly different set of contaminants, with the focus on those expected to be related to road and land runoff—PCBs, PBDEs, organochlorine compounds, PAHs, currently used pesticides, trace elements, mercury, and oil and grease. A complex mixture of compounds was detected in stormwater runoff, with detections of 114 (58 percent) of the 195 compounds analyzed. The detection patterns and concentrations measured in the stormwater-runoff samples, however, were more heterogeneous than in the WWTP-effluent samples. This reflects differences in various factors, including suspended-sediment concentrations and known contamination sources present in some watersheds. Trace elements and PAHs, which are related to automobiles and impervious surfaces, were the most widespread compound classes detected in stormwater runoff, a typical finding in stormwater runoff in urban areas.

With a better understanding of the presence of these contaminants in the environment, future work can focus on developing research to characterize the effects of these contaminants on aquatic life and prioritize toxic-reduction efforts for the Columbia River Basin.

Introduction

The Columbia River drains 259,000 square miles of the Pacific Northwest, and flows more than 1,200 miles from its headwaters in the Canadian Rockies of British Columbia. The river drains areas in Montana, Idaho, Washington, Oregon, and Wyoming, before flowing along the border of Washington and Oregon to its mouth at the Pacific Ocean. The rivers and streams of the Columbia River Basin carry the fourth largest volume of runoff in North America. The approximately 8 million people who live in the basin depend on its resources for their health and livelihood (Independent Scientific Advisory Board, 2007; U.S. Environmental Protection Agency, 2009a). Similarly, hundreds of fish and wildlife species, including 12 stocks of threatened and endangered salmonid species, rely on the ecosystem for their food sources, security, and habitat. Therefore, the Columbia River Basin is of environmental and cultural significance for all its inhabitants.

With growing scientific concern about the health of the ecosystem, efforts have been increased to make the public aware of the presence of toxic contaminants in the environment and the unknowns with regard to the potential adverse effects of these contaminants on the inhabitants of the ecosystem. Contaminants are chemicals introduced to the environment in amounts that can be harmful to fish, wildlife, or people. Many of these contaminants enter the environment through the production, use, and disposal of numerous chemicals that offer improvements in industry, agriculture, medical treatment, and common household conveniences.

Several studies have been completed throughout the Columbia River Basin in the past 10–20 years in an effort to characterize contaminant concentrations in water, sediment, and fish. In 2005, the U.S. Environmental Protection Agency (EPA) joined other Federal, State, Tribal, local, and nongovernmental organizations in forming the Columbia River Toxics Reduction Working Group in an effort to coordinate this work and share information. The goal of the group is to reduce toxics in the basin and prevent further contamination. In 2009, EPA produced the *State of the River Report for Toxics* to document the current knowledge in the basin with regard to certain classes of compounds, and to open communication for developing future solutions for addressing toxics reduction (U.S. Environmental Protection Agency, 2009a).

Through this process, the working group acknowledged that an adequate understanding of sources of these contaminants must precede development of efficient and effective toxic-reduction efforts. In a national survey of 139 streams in 1999–2000, organic wastewater contaminants were detected in 80 percent of the streams surveyed (Kolpin and others, 2002). Of the 95 contaminants analyzed, fecal steroids, insect repellants, caffeine, antimicrobial disinfectants, fire retardants, and nonionic detergent metabolites were commonly detected classes. In 2004–05, some of these same pharmaceuticals, antibiotics, and anthropogenic organic compounds (AOCs) were analyzed in samples collected from the main stem Columbia River near Portland and Longview (Lower Columbia River Estuary Partnership, 2007). Less than 20 contaminants were detected in these filtered-water samples at concentrations less than 1 microgram per liter (μg/L, parts-per-billion) range (Morace, 2006). This would suggest that the large volume of water flowing in the Columbia River dilutes the concentrations of these manmade contaminants. In contrast, when Nilsen and others (2007) analyzed surficial bed sediments in the lower Columbia River main stem and several tributaries, 49 different AOCs were detected, supporting the need to analyze multiple media when assessing contaminant issues. Nilsen and others (2007) detected endocrine-disrupting compounds (contaminants that block or mimic hormones in the

Interstate Highway 5 (I-5) bridge on the Columbia River from Hayden Island, Portland, Oregon, October 2009.

body and cause harm to fish and wildlife) at 22 of 23 sites sampled, with concentrations in the parts-per-billion range. The studies hint at the presence of these contaminants in the environment, but the extent of their presence throughout the basin is poorly understood.

Recent research has raised questions about potential effects on fish, shellfish, wildlife, and human health from even trace exposure to these contaminants, including chronic effects (Kidd and others, 2007; Ings and others, 2011), reproductive disruption (Vajda and others, 2008; Colman and others, 2009; Jenkins and others, 2009), and physiological changes (Hoy and others, 2011). Little is known, however, about the extent of the environmental occurrence, transport, and ultimate fate of these contaminants in the Columbia River ecosystem. To efficiently and effectively reduce loadings of these compounds to the river, sources and pathways of contaminants need to be identified. Numerous studies have shown that WWTP effluent and stormwater runoff contribute contaminants to their receiving waters (Boyd and others, 2004; Kolpin and others, 2004; Glassmeyer and others, 2005; Phillips and Chalmers, 2009). These two pathways act as integrators of human activities and offer an area where changes could be made to lessen their effects on the environment. This study focused on WWTP effluent and stormwater runoff to characterize how potential contaminants could be contributed through these pathways. A better understanding of the way contaminants enter the Columbia River Basin may help water managers reduce of the occurrence of contaminants in the basin.

Purpose and Scope

This report presents the results of a study to (1) assess contaminant concentrations directly contributed to the Columbia River through WWTP effluent and stormwater runoff from adjacent urban environments, (2) evaluate instantaneous loadings to the Columbia River Basin from inputs of this type, and (3) provide information to water managers and policy makers to help with decision making about future sampling efforts and reduction activities. The data from this study provide an initial assessment of a broad array of contaminants that to date have little information available on different sources in the Columbia River Basin. These data will be a useful first step to (1) identify the contaminants of highest interest, (2) indicate the most important sources of these contaminants, and (3) prioritize contaminant-reduction efforts.

This investigation resulted from a scientific and financial partnership between the USGS, Columbia River Inter-Tribal Fish Commission, Lower Columbia Estuary Partnership, Columbia Riverkeeper, and Northwest Environmental Defense Center, all agencies involved in the Columbia River Toxics Reduction Working Group.

Sampling Design and Methods

Sampling Sites

This study was designed to characterize WWTP effluent and stormwater runoff directly entering the Columbia River. These pathways were examined separately, however, by focusing specifically on what contaminants were of interest for each pathway. This study was not designed as a paired study to compare the differences in these pathways, but rather to characterize what could be contributed by each. Because this was a reconnaissance study, the cities where samples were collected were selected throughout the basin to provide a range in sampling location, population characteristics, and climate setting—in downstream order along the Columbia River, they include Wenatchee, Richland, Umatilla, The Dalles, Hood River, Portland, Vancouver, St. Helens, and Longview (fig. 1, table 1).

In each city, one sample was collected from the WWTP effluent just prior to where it enters the river (table 2). These one-time samples represented a variety of treatment techniques varying by the size of the treatment plant and the type of disinfection used. This study was not designed to evaluate these treatment techniques or differentiate the associated concentrations, but rather to collect preliminary data. A stormwater-runoff sample also was collected directly from a pipe in each city just prior to where the runoff enters the receiving waters, except for in Umatilla, Oregon, where the stormwater flowed into a percolation field (table 3). In the Portland/Vancouver area (fig. 2), extra samples were collected—two locations in Vancouver, Washington, and two locations in Portland, Oregon, where stormwater enters the Columbia River. Rather than draining directly to the Columbia River, much of the stormwater from the eastern Portland area is delivered to the Columbia Slough (fig. 2) which flows through Portland and enters the Willamette River just before it converges with the Columbia River. Much of the remaining stormwater from the western and southern areas of Portland flows through pipes into the Willamette River. For this study, stormwater runoff from four pipes discharging to the Willamette River also was sampled.

Table 1. Precipitation, population, age, and income information for cities where samples were collected, Columbia River Basin, Washington and Oregon.

[Data from U.S. Census Bureau (2010). **Symbols:** <, less than; >, greater than]

City	State	County	Average annual precipitation (inches)	Population	Population density (people per square mile)	<18 years	18–65 years	>65 years	Median age (years)	Median annual household income (U.S. dollars)	Average household size (persons)
Wenatchee	Washington	Chelan	9	31,925	4,110	25	61	14	36.4	$42,600	2.55
Richland	Washington	Benton	7	48,058	1,350	25	62	13	38.4	$62,200	2.53
Umatilla	Oregon	Umatilla	8	6,906	1,560	23	73	4	31.5	$46,800	3.08
The Dalles	Oregon	Wasco	14	13,620	2,150	25	57	18	39.3	$44,100	2.40
Hood River	Oregon	Hood River	32	7,167	2,810	28	58	14	34.9	$47,200	2.43
Portland	Oregon	Multnomah	37	583,776	4,380	20	70	10	35.5	$48,100	2.27
Vancouver	Washington	Clark	42	161,791	3,480	24	64	12	35.6	$48,000	2.46
St. Helens	Oregon	Columbia	46	12,883	2,840	28	64	8	33.3	$53,500	2.74
Longview	Washington	Cowlitz	48	36,648	2,530	24	60	16	39.1	$39,000	2.32

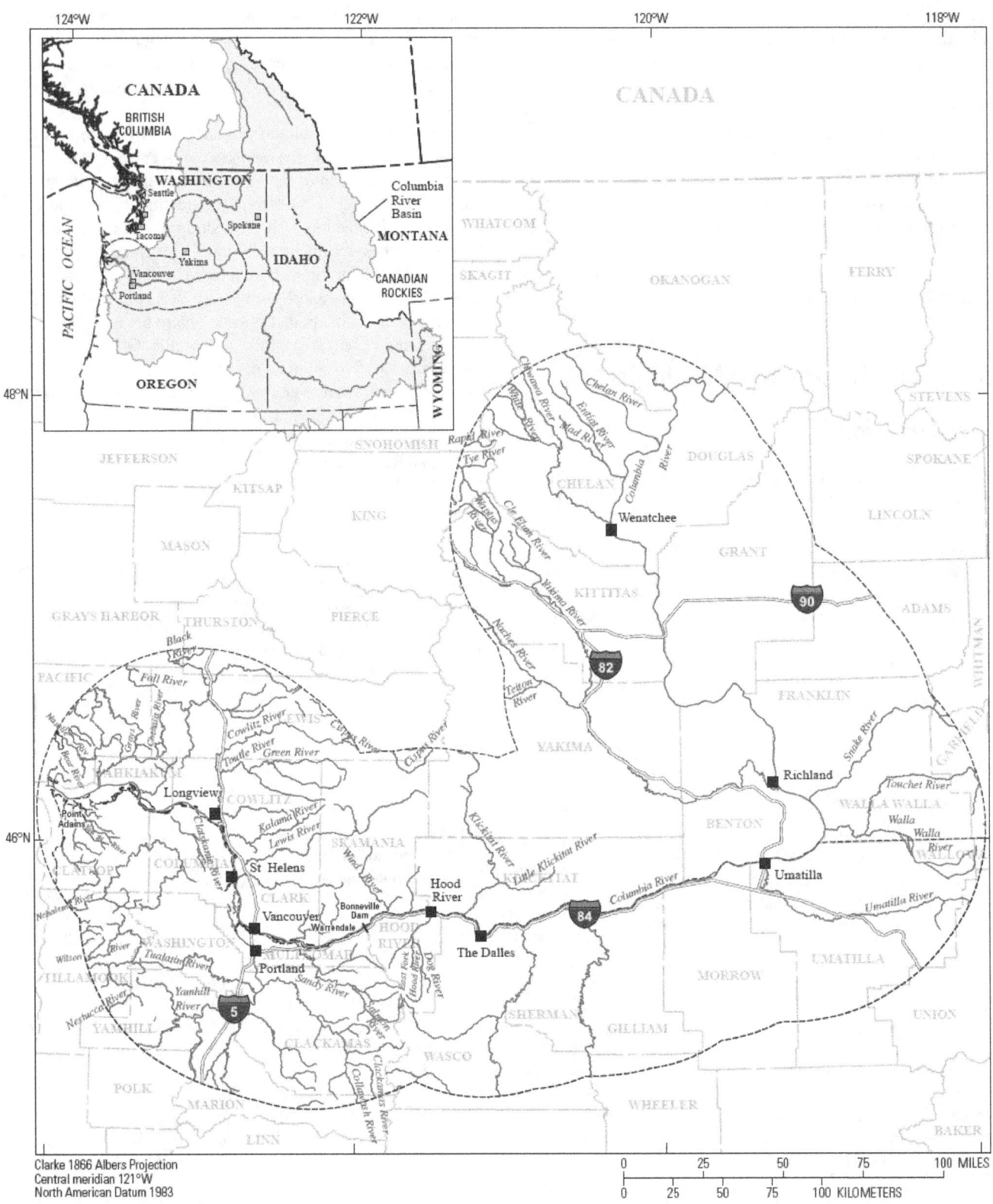

Figure 1. Cities where samples were collected from wastewater-treatment plants and stormwater runoff in the Columbia River Basin, Washington and Oregon, 2008–10.

Table 2. Descriptions of wastewater-treatment plants where samples were collected, Columbia River Basin, Washington and Oregon, 2008–09.

[Cities shown in downstream order. **Abbreviations:** Mgal/d, million gallons per day; 7Q10, the lowest streamflow for 7 consecutive days that occurs on average once every 10 years; ft³/s, cubic foot per second; UV, ultraviolet; WWTP, wastewater treatment plant; NA, not available]

City	Station name	Point of discharge, Columbia River mile	Permit design flow (Mgal/d)	Columbia River 7Q10 streamflow (ft³/s)	Wastewater-treatment plant description
Wenatchee	City of Wenatchee WWTP effluent at Wenatchee, Wash.	466.6	7.1	51,557	Activated sludge plant, secondary-level treatment and UV disinfection.
Richland	City of Richland WWTP effluent at Richland, Wash.	337.1	11.4	52,700	Semi-plug flow conventional activated sludge, secondary clarification, and chlorination.
Umatilla	City of Umatilla WWTP effluent at Umatilla, Oreg.	289	0.92	NA	Oxidation ditch with UV disinfection.
The Dalles	City of The Dalles WWTP effluent at The Dalles, Oreg.	189.5	4.15	80,637	Activated sludge plant with UV disinfection.
Hood River	City of Hood River WWTP effluent at Hood River, Oreg.	165	2	74,000	Activated sludge plant with UV disinfection.
Portland	Columbia Blvd WWTP effluent at Hayden Island, Oreg.	105.5	72	79,436	Conventional activated sludge, secondary clarification, chlorine disinfection.
Vancouver	Vancouver Westside WWTP effluent at Vancouver, Wash.	105	28	79,436	Industrial Pretreatment Lagoon, secondary activated sludge, UV disinfection, sludge incineration.
St. Helens	City of St. Helens WWTP effluent at St. Helens, Oreg.	86.9	45	88,900	Combined municipal and kraft mill aerated stabilization basin.
Longview	Three Rivers Regional WWTP effluent at Longview, Wash.	67.5	26	97,400	Conventional activated sludge, secondary clarification, chlorine disinfection, dechlorination.

Table 3. Stormwater-runoff sampling locations in the Columbia River Basin, Washington and Oregon, 2009–10.

[Outfalls are shown in downstream order]

Short name	Outfall station name	County	Station No.	Date	Time
Wenatchee	Wenatchee stormwater outfall at Chehalis Street near footbridge, Wash.	Chelan	472506120180900	12-21-09	1340
Richland	Richland stormwater outfall near Columbia Park West, Wash.	Benton	461414119125400	05-02-09	1200
Umatilla	Umatilla stormwater outfall south end of percolation field, Oreg.	Umatilla	455448119205900	10-04-09	0920
The Dalles	The Dalles stormwater outfall at Klindt Point, Oreg.	Wasco	453750121115300	02-23-09	1210
Hood River	Hood River stormwater outfall near Nichols Basin, Oreg.	Hood River	454256121304100	02-23-09	1310
Portland1	Stormwater outfall near I-205 bridge at northeast 112th Avenue, Oreg.	Multnomah	453424122324400	10-14-09	1100
Vancouver1	Stormwater outfall near southeast corner of Fort Vancouver, Wash.	Clark	453705122393300	12-16-09	1340
Vancouver2	Stormwater outfall under I-5 bridge at Vancouver, Wash.	Clark	453717122402400	12-16-09	1210
Portland2	Stormwater outfall under I-5 bridge on Hayden Island, Oreg.	Multnomah	453651122403900	10-26-09	1210
Willamette1	Stormwater outfall downstream of west end of Marquam Bridge, Oreg.	Multnomah	453025122401700	06-04-10	0840
Willamette2–Dec.	Stormwater outfall under west end of St. John's railroad bridge, Oreg.	Multnomah	453431122445800	12-15-09	1330
Willamette2–May	Stormwater outfall under west end of St. John's railroad bridge, Oreg.	Multnomah	453431122445800	05-26-10	1310
Willamette3	Stormwater outfall downstream of west end of St. John's railroad bridge, Oreg.	Multnomah	453431122445900	12-15-09	1310
Willamette4	Stormwater outfall west end North Ramsey Boulevard, Oreg.	Multnomah	453726122471500	05-26-10	1410
St Helens	St Helens stormwater outfall at boat launch on River Street, Oreg.	Columbia	455203122475600	03-30-10	1310
Longview	Longview stormwater ditch at 99 Oregon Way, Wash.	Cowlitz	460703122570000	03-30-10	1410

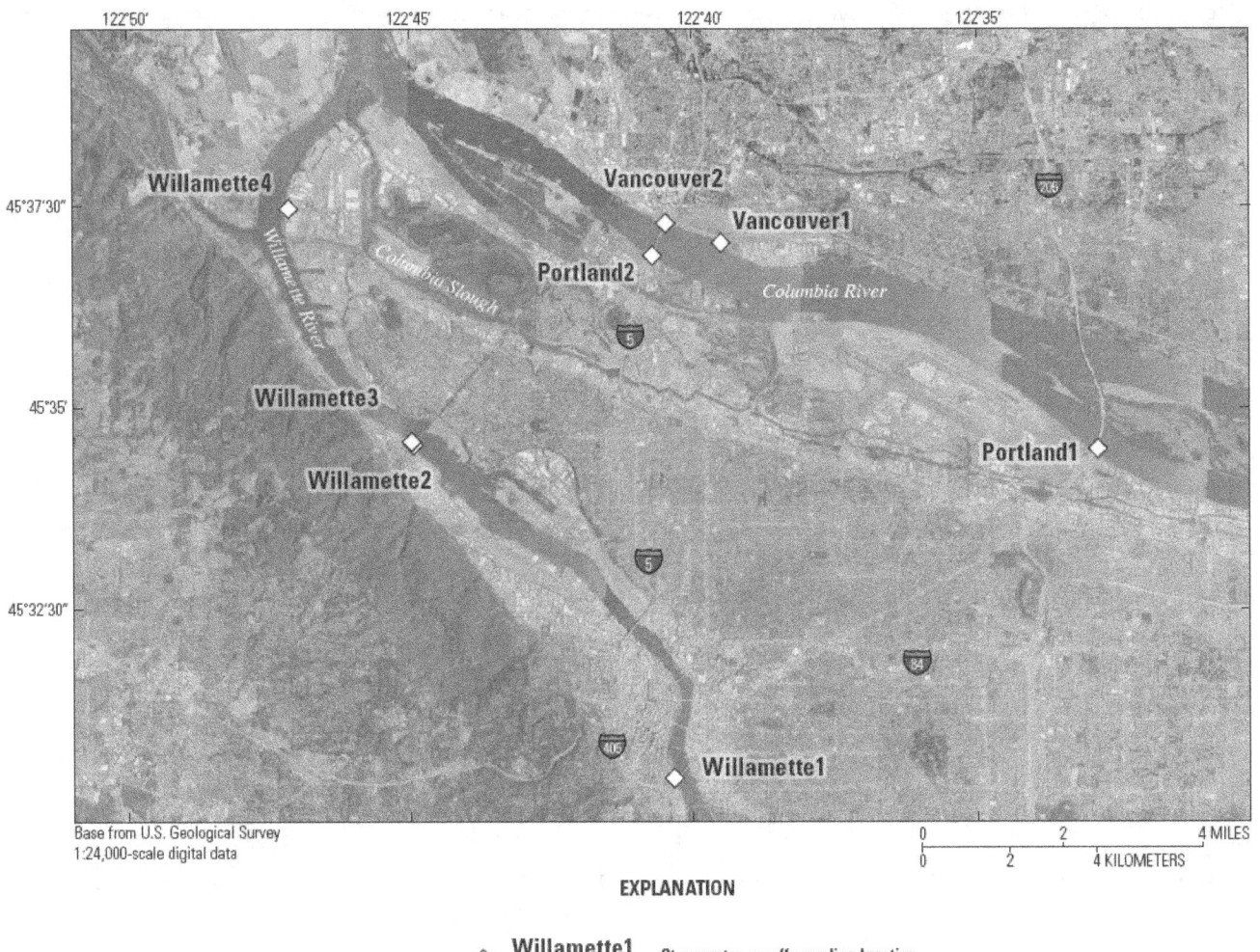

Figure 2.   Selected stormwater-runoff sampling locations in the Portland, Oregon, and Vancouver, Washington, area, Columbia River Basin, 2009–10.

Many cities, and Portland in particular, have older sewer systems that mix untreated sewage and stormwater runoff. When it rains, these systems are overwhelmed and combined sewer overflows (CSOs) carry untreated sewage to the receiving waters. During wet weather, Portland's combined sewers overflow into the Willamette River an average of 100 times per year (Portland Bureau of Environmental Services, 2010). Phillips and Chalmers (2009) have shown that untreated discharge from CSOs can be an important source of contaminants to receiving waters. In an effort to prevent these CSOs and improve water quality, the city of Portland has constructed several "big pipes"—the Columbia Slough Big Pipe (completed in 2000), the West Side Big Pipe (completed

in 2006), and the East Side Big Pipe (completed in 2011) (Portland Bureau of Environmental Services, 2010). These large (12- to 22-ft diameter) pipes help store and transport the overflow so that it can be treated before it is discharged. Combined sewer overflows to the Willamette River will be reduced by 94 percent when all east-side CSO construction is complete. These Big Pipes influenced the sampling locations for this study because they prevent stormwater pipes from delivering runoff to locations that previously received the runoff. Sites selected on the west side of the Willamette River are upstream (Willamette1) and downstream (Willamette2 and Willamette3) of the Big Pipe drainage areas (fig. 2).

Sampling and Analytical Methods

To characterize the nature of the water entering the Columbia River, each sample was collected in the WWTP at a point in the effluent stream past any treatment and just before the effluent enters the river. This was a dip sample in the effluent stream at most WWTPs, but in Longview, St. Helens, and Portland, the samples were pumped into the bottles by the onsite pumps. One sample was collected at each of the nine cities, except at Portland where samples were collected three times throughout the day (9 a.m., 12 p m., and 3 p m.) to examine temporal variability (table 4). Therefore, 11 WWTP-effluent samples were collected.

Similar to the samples collected at WWTPs, the stormwater samples were collected from the end of the pipe just before it entered the river. One sample was collected at each of the nine cities, except in Portland and Vancouver, where two locations were sampled in each city. An additional four stormwater locations were sampled along the Willamette River in an effort to better characterize stormwater runoff in the Portland area. Thus, a total of 15 locations were sampled for stormwater runoff.

The original project plan did not include analysis of currently used pesticides in WWTP-effluent samples. Because pesticides make up 34 percent of the persistent pollutants list in Oregon Senate Bill (SB) 737 (Oregon Department of Environmental Quality, 2010a), however, a decision was made to revisit each WWTP during December 2009 to collect samples of effluent for analysis of pesticides and mercury. Oregon Department of Environmental Quality (ODEQ) was required by SB 737 to develop a list of priority persistent bioaccumulative toxics (persistent pollutants) that have a documented effect on human health, wildlife, and aquatic life. The 52 largest municipal WWTPs in Oregon analyzed their effluent in July and November 2010 for these persistent pollutants, and they currently are developing reduction plans for those compounds that were detected above plan initiation levels (PILs) determined as part of this process.

All samples for this study were collected using standard methods described by U.S. Geological Survey (USGS; variously dated). Samples were placed into glass or Teflon® bottles depending on the type of analysis, and then composited into either a glass carboy or Teflon churn for processing. Samples were collected at most of the WWTPs in December 2008, but samples were collected in

Stormwater-runoff sample in 20-liter glass carboy, collected from a pipe under the I-5 bridge on Hayden Island, Oregon, October 2009.

St. Helens and Longview in December 2009 because they were added later in the project (table 4). Wenatchee was resampled in 2009 due to sampling errors with the filtration apparatus in 2008 that compromised some of the analyses. The stormwater samples were collected throughout spring and winter storms of 2009 and 2010.

Samples were placed on ice until they could be processed and shipped to the appropriate laboratory; most samples were shipped in less than 3 hours. Volunteers were used to collect stormwater samples in remote locations. These samples were shipped to the USGS Oregon Water Science Center before they were processed, resulting in a holding time of about 40 hours or less. While the sample was being mixed (required to resuspend any settled solids), unfiltered-water samples were drained into their respective bottles. For filtered-water analyses, aliquots of the sample were filtered through a 142-mm diameter, 0.7-μm pore-size glass-fiber filter and collected into amber glass bottles to be sent to the USGS National Water-Quality Laboratory (NWQL) in Denver, Colorado.

Table 4.   Summary of sampling activities, Columbia River Basin, Washington and Oregon, 2008–10.

[Station names are shown in tables 2 and 3. Constituents analyzed, reporting limits, and method parameters are presented in appendix A]

City or short name	Date	Time	Suspended sediment	Anthropogenic organic compounds in unfiltered water (table A2)	Pharmaceuticals in filtered water (table A3)	Halogenated compounds on solids filtered from samples (table A1)	Currently used pesticides in filtered water (table A4)	Mercury and methylmercury in unfiltered water	Polycyclic aromatic hydrocarbons in unfiltered water (table A5)	Trace elements in unfiltered and filtered water (table A6)	Oil and grease in unfiltered water
Wastewater-treatment-plant effluent samples											
2008 visit to initial seven cities											
Wenatchee	12-02-08	1010	X	X	X	X					
Richland	12-04-08	0900	X	X	X	X					
Umatilla	12-03-08	0840	X	X	X	X					
The Dalles	12-05-08	0830	X	X	X	X					
Hood River	12-10-08	0950	X	X	X	X					
Portland (a.m.)	12-09-08	0900	X	X	X	X					
(noon)	12-09-08	1150	X	X	X	X					
(p.m.)	12-09-08	1500	X	X	X	X					
Vancouver	12-08-08	0940	X	X	X	X					
2009 revisit and additional cities											
Wenatchee	12-01-09	0850	X	X	X	X	X	X			
Richland	12-02-09	0820					X	X			
Umatilla	12-02-09	0950					X	X			
The Dalles	12-02-09	1200					X	X			
Hood River	12-02-09	1310					X	X			
Portland	12-10-09	0840					X	X			
Vancouver	12-02-09	1510					X	X			
St. Helens	12-03-09	0900	X	X	X	X	X	X			
Longview	12-08-09	0810	X	X	X	X	X	X			
Stormwater-runoff samples											
2009 and 2010 storms											
Wenatchee	12-21-09	1340	X			X	X	X	X	X	X
Richland	05-02-09	1200	X			X	X		X	X	X
Umatilla	10-04-09	0920	X			X	X		X	X	X
The Dalles	02-23-09	1210	X			X	X		X	X	X
Hood River	02-23-09	1310	X			X	X		X	X	X
Portland1	10-14-09	1100	X			X	X		X	X	X
Portland2	10-26-09	1210	X			X	X		X	X	X
Vancouver1	12-16-09	1340	X			X	X	X	X	X	X
Vancouver2	12-16-09	1210	X			X	X	X	X	X	X
Willamette1	06-04-10	0840	X			X	X	X	X	X	X
Willamette2–Dec	12-15-09	1330	X			X	X	X	X	X	X
Willamette2–May	05-26-10	1310	X			X	X	X	X	X	X
Willamette3	12-15-09	1310	X			X	X	X	X	X	X
Willamette4	05-26-10	1410	X			X	X	X	X	X	X
St. Helens	03-30-10	1310	X			X	X	X	X	X	X
Longview	03-30-10	1410	X			X	X	X	X	X	X

Analytical Methods for Wastewater-Treatment-Plant-Effluent Samples

A full listing of all constituents analyzed, reporting limits, and method numbers is presented in appendix A. Because halogenated compounds like flame retardants, PCBs, and certain pesticides (table A1) are hydrophobic (preferentially associated with sediment particles), solid samples were collected for analysis. Because WWTP effluent is low in solids by design, about 20 liters (L) of effluent were filtered for each WWTP. These filters were sent to the NWQL for the analysis of halogenated compounds on the filtered solids. These analyses were done as an adaptation to the method used for analyzing these compounds in sediments (Steven Zaugg, National Water-Quality Laboratory, written commun., March 16, 2010), which involved extracting all material collected on the filters and concentrating it down to 1 mL of extract, which was analyzed for the entire suite of halogenated compounds. These concentrations, therefore, provide a measure of the hydrophobic compounds detected in the particulate phase and do not account for compounds present in the dissolved phase.

For the WWTP-effluent samples, AOCs in unfiltered water (table A2) were analyzed at the NWQL by continuous liquid-liquid extraction and gas chromatography/mass spectrometry (GC/MS) using methods described by Zaugg and others (2006). Human-health pharmaceuticals (table A3) and currently used pesticides (table A4) in filtered-water samples were analyzed at NWQL by GC/MS using methods detailed by Zaugg and others (1995), Lindley and others (1996), Sandstrom and others (2001), Madsen and others (2003), and Furlong and others (2008). In 2009 and 2010, unfiltered water samples were preserved with hydrochloric acid and sent to the USGS Wisconsin Mercury Research Laboratory for the analysis of total mercury and methylmercury by methods described by U.S. Environmental Protection Agency (2002) and DeWild and others (2002), respectively. Suspended-sediment concentrations were determined at the Cascades Volcano Observatory Sediment Laboratory in Vancouver, Washington, according to methods detailed by Guy (1969).

For samples collected at each city in 2008, a 1-L unfiltered WWTP-effluent sample was shipped on ice to the Columbia Environmental Research Center (CERC) in Columbia, Missouri,

Filtering wastewater-treatment-plant effluent from the City of Portland, Oregon, December 2008.

to be screened for total estrogenicity using the yeast estrogen screen (YES) by methods described by David Alvarez (U.S. Geological Survey, written commun., November 6, 2009). The YES assay uses recombinant yeast cells with a human estrogen receptor. If these cells bind to an estrogen or estrogen-mimic in the sample, then a number of biochemical reactions occur and result in a color change (Routledge and Sumpter, 1996; Rastall and others, 2004). This color change can be measured spectrophotometrically and the estradiol equivalent factor (EEQ) for the sample can be determined. The EEQ is an estimate of the amount of 17β-estradiol, a natural hormone, which would be needed to give an equivalent response to that of the complex mixture of chemicals present in the sample (Rastall and others, 2004; Alvarez and others, 2008). Therefore, a higher measured EEQ indicates a higher estrogenicity of the sample.

Analytical Methods for Stormwater-Runoff Samples

Halogenated compounds on solids, currently used pesticides in filtered water, mercury and methylmercury in unfiltered water, and suspended-sediment samples collected from the stormwater runoff were processed in the same way as described for the WWTP-effluent samples. Additionally, stormwater-runoff samples were collected for the analysis of PAHs in unfiltered water (table A5) and trace elements in both filtered and unfiltered water (table A6). These samples were analyzed at the NWQL by methods described by Fishman and Friedman, (1989), Fishman (1993), Hoffman and others (1996), Garbarino and Struveski (1998), Garbarino and Damrau (2001), and Garbarino and others (2006). Unfiltered-water samples were subsampled into bottles with sulfuric acid preservative and shipped on ice to the TestAmerica Laboratory in Arvada, Colo., for analysis of oil and grease by EPA method 1664A (U.S. Environmental Protection Agency, 1999).

Reporting of Data

When an analyte is measured in a laboratory, it is either detected or not detected. When it is not detected, it is reported as "censored" or less than the reporting limit (RL). This does not mean that the analyte is not present; it simply means that it could not be detected in a sample under the conditions present in the laboratory or the sample matrix. The analyte may be present, but at a concentration lower than the instrument can measure. Likewise, the presence of other material or analytes in the sample may be causing interference, preventing the accurate quantification of the analyte in the sample, or the analyte may not be present at all. If, however, the analyte is detected, it may be reported in several different ways. If it is detected at a concentration greater than the RL, then the value is simply reported at the concentration measured. If the analyte is a "poor performer" (long-term variability or poor recovery) in laboratory performance samples or if matrix problems caused interference for that analyte in the sample, the measured concentration may be qualified as an estimated (E) value.

The concentration also may be reported as an estimated value if the analyte is detected at a concentration less than the RL but greater than the method detection limit (MDL). The MDL is a statistically derived minimum concentration that can be measured with a 99 percent confidence of being greater than zero (Oblinger Childress and

Filter paper after filtering stormwater-runoff sample from the City of Umatilla, Oregon, October 2009.

others, 1999; Bonn, 2008). Therefore, there is a less than a 1 percent chance that an analyte will be reported as a false positive, or that the concentration was reported but the analyte was not present. If the analyte is detected at a concentration less than the MDL or RL (for those analytes for which a MDL has not yet been established), then, in this report, the result is reported as "Present," indicating that the presence of the analyte was verified, but that the concentration was too small to be quantified. The NWQL reevaluates the RL and MDL values every year and adjusts them as needed based on the laboratory performance data. Because of these adjustments, multiple RLs may be shown for a given analyte. Additionally, matrix interference issues, which were numerous in this study due to the complex nature of the effluent and runoff, can cause the RL for a certain compound for individual samples to be raised as well.

The data for the halogenated compounds on solids were reported from the NWQL as the mass of the given analyte detected in the sample (in nanograms). This mass was then divided by the number of liters filtered, to obtain a concentration of the analyte for the sample (in nanograms per liter, ng/L). Detections for this analysis were reported only if the mass was greater than the RL or five times the highest value reported in the laboratory blank, trip blank, or analyses of the filter papers themselves. Detections less than these levels were reported as "Present."

Quality Assurance

Quality assurance is the analysis of quality-control (QC) data as a means to assess potential contamination and variability associated with sampling and laboratory techniques. Quality control samples for this study comprised field blanks and replicate environmental samples (table 5), as well as internal laboratory QC data such as set blanks, set spikes, and surrogate recoveries. Between 1 and 3 blanks and 2 and 4 replicates were collected for each analytical method. For some

Table 5. Summary of quality-control (QC) analyses performed for this study, Columbia River Basin, Washington and Oregon, 2008–10.

[Station names are shown in tables 2 and 3]

City or short name	Date	Type of quality-control sample	Suspended sediment	Anthropogenic organic compounds in unfiltered water	Pharmaceuticals in filtered water	Halogenated compounds on solids filtered from samples	Currently used pesticides in filtered water	Mercury and methylmercury in unfiltered water	Polycyclic aromatic hydrocarbons in unfiltered water	Trace elements in unfiltered and filtered water	Oil and grease in unfiltered water
Wastewater-treatment-plant effluent samples											
2008 visit to initial seven cities											
Umatilla	12-03-08	Blank	X	X	X	X[1]					
Vancouver	12-08-08	Replicate	X	X	X	X					
2009 revisit and additional cities											
Umatilla	12-02-09	Blank						X			
Vancouver	12-02-09	Replicate					X	X			
St Helens	12-03-09	Replicate	X	X	X	X	X	X			
Stormwater-runoff samples											
2009 and 2010 storms											
Wenatchee	12-21-09	Replicate	X			X	X	X	X	X	X
The Dalles	02-23-09	Blank	X				X		X	X	X
Portland2	10-26-09	Replicate	X			X	X		X	X	X
Willamette1	06-07-10	Blank	X			X	X	X	X	X	X

[1] Filter papers were analyzed instead of a field blank for this method.

combinations of method and sample type (currently used pesticides in WWTP samples), no blanks were collected, although for most combinations, one blank and two replicates were collected. QC samples were collected throughout the sampling periods to assess any annual variability in laboratory performance. Results of all of these QC samples were used to qualify the environmental data.

Field blanks were collected by passing a volume of contaminant-free water (organic blank water) through sampling and processing equipment that an environmental sample would contact. The results of field blanks are used to assess contamination issues associated with cleaning, sampling, processing, or transporting the sample. In addition to a field blank for the halogenated compounds on solids, the filter papers themselves were run through the process to assess whether they may be affecting the analysis.

Replicate environmental samples test for precision, which is a measure of the variability between two or more samples caused by variability in laboratory processing techniques and measurement precision. Replicate samples were collected consecutively, except for the stormwater sample collected at Portland2. That sample was collected into one glass carboy, agitated to resuspend solids, and then split in the laboratory during processing.

"Surrogate compounds" have properties similar to those of the target compounds; surrogate compounds are added to the sample at the laboratory and analyzed as part of the list of analytes. Surrogate compounds are expected to behave similarly to the target analytes and are used to monitor the performance of the method used for the target analytes they represent. The NWQL uses the surrogate recoveries to assess problems associated with individual samples or sets of samples, but also uses long-term surrogate recoveries to assess long-term analytical precision. Surrogate recoveries in this study were good for blank samples but generally low for environmental samples (table 6). This was probably due to matrix-interference issues. The actual concentrations in the samples may have been underestimated by the analyses; therefore, this report represents a conservative measure of the contaminants delivered by WWTP effluent and stormwater runoff. Because of the large variations in sample recoveries and sample performance, care should be used in drawing comparisons between sample sets. Although the validity of quantitative comparisons may be compromised by this variability, qualitative analyses, based on the presence or absence of these compounds, provide a way to compare these types of datasets.

Table 6. Summary of surrogate recoveries, Columbia River Basin, Washington and Oregon, 2008–10.

[Station names are shown in tables 2 and 3. Values reported in percent. **Abbreviations:** ND, not determined because of matrix interferences or sample preparation errors; E, estimated; HCH, hexachlorocyclohexane; DDT, dichlorodiphenyltrichloroethane; PCB, polychlorinated biphenyl; NA, not analyzed; currently used pesticides were not analyzed in 2008]

Surrogate analyte	Wenatchee (2008)	Wenatchee (2009)	Richland	Umatilla	The Dalles	Hood River	Portland (a.m.)	Portland (noon)	Portland (p.m.)	Portland (2009)	Vancouver	St. Helens	Longview	Blank
Anthropogenic organic compounds in unfiltered water														
Bisphenol A-d3	87	89	38	94	0	18	31	37	ND	NA	0	94	89	88
Caffeine-c13	84	65	21	97	66	79	91	94	62	NA	36	80	75	89
Decafluoro-biphenyl	58	56	26	84	59	66	56	61	48	NA	30	51	61	76
Fluoranthene-d10	65	67	37	90	61	75	62	64	47	NA	28	63	68	89
Pharmaceuticals in filtered water														
Carbamaz-epine-d10	17	22	27	20	13	23	11	8	10	NA	12	29	15	106
Ethyl nicotinate-d4	E 138	65	E 100	E 90	E 101	114	73	81	53	NA	83	54	39	E 134
Currently used pesticides in filtered water														
Diazinon-d10	NA	138	E 109	E 105	E 104	E 113	NA	NA	NA	132	E 108	117	116	NA
alpha-HCH-d6	NA	ND	ND	ND	ND	ND	NA	NA	NA	81	ND	102	88	NA
Halogenated compounds on solids														
DDT-d8	NA	5	24	42	31	15	35	20	54	NA	18	12	34	64
Dibromo-octafluoro-biphenyl	NA	5	61	63	62	54	65	34	66	NA	56	12	25	53
PCB-202-13C12	NA	4	34	53	43	35	31	22	50	NA	32	7	13	67

Table 6. Summary of surrogate recoveries, Columbia River Basin, Washington and Oregon, 2008–10.—Continued

[Station names are shown in tables 2 and 3. Values reported in percent. **Abbreviations:** ND, not determined because of matrix interferences or sample preparation errors; E, estimated; HCH, hexachlorocyclohexane; DDT, dichlorodiphenyltrichloroethane; PCB, polychlorinated biphenyl; NA, not analyzed, for example, currently used pesticides were not analyzed in 2008]

Surrogate analyte	Wenatchee	Richland	Umatilla	The Dalles	Hood River	Portland1	Vancouver1	Vancouver2	Portland2	Willamette1	Willamette2 -Dec	Willamette2 -May	Willamette3	Willamette4	St. Helens	Longview	Blank– Hood River	Blank– Willamette1
							Stormwater-runoff samples											
							Currently used pesticides in filtered water											
Diazinon-d10	114	E 153	138	134	125	153	108	105	134	116	126	125	109	116	98	120	100	103
alpha-HCH-d6	80	91	79	97	93	84	92	90	105	92	81	102	91	96	99	99	87	96
							Polycyclic aromatic hydrocarbons (PAHs) in unfiltered water											
2-Fluoro-biphenyl	70	84	63	52	93	63	73	73	46	80	73	64	77	63	33	67	87	89
2-Fluoro-phenol	66	78	71	23	37	62	64	61	28	79	59	57	74	43	20	54	54	49
Nitrobenzene-d5	103	94	97	57	97	94	85	88	53	102	88	83	90	74	38	79	92	95
Phenol-d5	52	57	58	9	32	58	52	48	28	60	49	57	65	31	19	46	45	37
Terphenyl-d14	45	64	37	34	84	35	55	45	30	49	55	34	52	41	34	52	88	87
2,4,6-Tribro-mophenol	91	113	94	38	47	93	90	93	61	108	102	95	103	73	23	76	63	54
							Halogenated compounds on solids											
DDT-d8	75	43	46	29	54	14	3	4	74	83	9	83	7	81	97	33	NA	82
Dibromo-octafluoro-biphenyl	62	82	25	45	29	20	3	4	30	52	7	66	7	38	104	23	NA	77
PCB-202-13C12	65	47	19	34	51	7	1	2	40	67	6	65	5	70	65	21	NA	53

Results of Quality-Control Data

There were only a few analytes with detections in the field blanks (table 7), and most were not at levels that warrant concern with respect to the environmental detections. For those compounds with detections in a blank sample (a field, filter, or set blank), the highest detected value in the blank was multiplied by five and this new value was used as a "raised reporting level." If a detected concentration was less than this raised reporting level, then it was reported as "Present," rather than the actual concentration.

Table 7. Summary of detections in blank samples, Columbia River Basin, Washington and Oregon, 2008–10.

[No detections in blanks for pharmaceuticals. Raised reporting limits were used to further qualify the environmental data—detections less than raised reporting limit have been reported as "Present." **Abbreviations:** E, estimated; PBDE, polybrominated diphenyl ether; PCB, polychlorinated biphenyl]

Compound detected in blank	Highest value in blank	Raised reporting limit (5 * blank value)	Lowest detected concentration
Anthropogenic organic compounds in unfiltered water, in micrograms per liter (field blank)			
Cholesterol	E 0.62	3.1	E 0.8
3-*beta*-Coprostanol	E 0.58	2.9	E 0.63
4-Nonylphenol monoethoxylate (sum of all isomers) [NP1EO]	E 0.22	1.1	E 0.32
para-Nonylphenol (total)	E 0.42	2.1	E 0.4
Phenol	E 0.19	0.95	E 0.18
Halogenated compounds on solids, in nanograms (field, filter, and set blanks)			
Dechlorane plus	0.21	1.1	0.19
Hexachlorobenzene (HCB)	0.22	1.1	1.0
PBDE-47	2.7	14	22
PBDE-99	0.84	4.2	2
PBDE-100	0.25	1.3	0.54
PCB-194	0.11	0.55	0.06
Pentachloroanisole	0.11	0.55	0.09
Triclosan	8.4	42	6.2
Polycyclic aromatic hydrocarbons (PAHs), in micrograms per liter (field blanks)			
1,4-Dichlorobenzene	E 0.014	0.07	E 0.015
Diethyl phthalate	E 0.155	0.78	E 0.22
Dimethyl phthalate	E 0.029	0.15	E 0.016
Trace elements in unfiltered and filtered water, in micrograms per liter (field blanks)			
Lead in unfiltered water	E 0.053	0.27	0.18
Chromium in filtered water	E 0.084	0.42	E 0.075
Mercury and methylmercury in unfiltered water, in nanograms per liter (field blanks)			
Total mercury in unfiltered water	0.24	1.2	1.9

Method blanks for the oil and grease analyses showed consistent and significant detections. Nearly one-half of the method-blank samples had detectable concentrations equal to at least one-half of the coinciding environmental-sample concentration. The environmental and blank results are reported together so that the user is aware of these issues (see section "Oil and Grease").

When comparing differences in concentrations from different sites or different times, the analytical and environmental variability must be considered. Examining environmental replicate data can help quantify this variability. Relative percent difference (RPD) values, which provide a measure of how well the concentrations from two samples agree, were calculated for all environmental replicate

data pairs (table 8). The RPD is calculated as the absolute difference between two values, normalized to the average value, and expressed as a percentage.

$$RPD \equiv \left| \frac{(Value1 - Value2)}{(Value1 + Value2)/2} \right| \times 100.$$

An RPD close to zero shows good agreement between the sample results, but some RPDs in this study are high. This probably is due to a number of variables, including methods used to analyze unfiltered water, low concentrations in samples, and the inherent variability in some of the methods, which may be linked to matrix effects.

Table 8. Summary of relative percent differences for replicate samples, Columbia River Basin, Washington and Oregon, 2008–10.

[Each analytical schedule and sample type had two replicate pairs except for mercury in stormwater runoff, which had only one replicate pair. **Abbreviations:** LC, laboratory code; SH, schedule; Storm, stormwater runoff; WWTP, wastewater-treatment-plant effluent; –, not applicable]

Analytical No.	Compound group description	Sample type	Number of compounds analyzed	Number of compounds detected in replicate pair #1	Number of compounds detected in replicate pair #2	Relative percent difference		
						Minimum	Median	Maximum
SH 4433	Anthropogenic organic compounds	WWTP	69	20, 22	34, 33	0	39	101
SH 2080	Pharmaceuticals	WWTP	14	6, 7	7, 7	0	13	74
LC 8093	Halogenated compounds	WWTP	60	19, 20	17, 23	0	23	56
LC 8093	Halogenated compounds	Storm	60	3, 1	15, 9	0	42	88
SH 2033	Currently used pesticides	WWTP	83	6, 6	6, 6	0	4	30
SH 2033	Currently used pesticides	Storm	83	7, 7	2, 3	0	6	21
–	Mercury and methylmercury	WWTP	2	2, 2	2, 2	2	25	40
–	Mercury and methylmercury	Storm	2	2, 2	–	2	–	12
SH 1383	Polycyclic aromatic hydrocarbons	Storm	56	8, 8	22, 25	0	17	58
SH 1264	Trace elements in filtered water	Storm	10	8, 8	8, 8	0	4	92
SH 1264	Trace elements in unfiltered water	Storm	10	7, 7	9, 8	0	4	92

Compound Classes

Several compound classes were analyzed in this study. A brief description of each is provided here to give an overview of what type of contaminants are in each class, potential sources of each class, and documented effects of some of these contaminants.

Anthropogenic Organic Compounds

Anthropogenic organic compounds are a broad suite of compounds that are typically associated with human, industrial, and agricultural wastewater and include detergent metabolites, flame retardants, personal care products, pesticides, plasticizers, PAHs, steroids, and other miscellaneous compounds. Although these compounds are associated with wastewater, it is important to note that the WWTP is not the source of these compounds, but simply a pathway by which these compounds can reach the ecosystem from urban environments.

Besides WWTP effluent, these compounds also can reach streams from runoff from land applications, industrial facilities, animal feed lots, and septic systems. In 1999–2000, 139 streams were sampled in 30 states across the United States, and AOCs were detected in 80 percent of the streams sampled with as many as 38 compounds in one sample (Kolpin and others, 2002). Some of these compounds bioaccumulate in biota and many are suspected or known endocrine disruptors, meaning they mimic hormones and can cause problems with the endocrine system, which affects reproduction and growth. Detergent metabolites nonylphenol and fragrances like tonalide and galaxolide have been shown to cause endocrine issues in fish (Schreurs and others, 2004), and the antimicrobial disinfectant triclosan can result in reduced algal diversity (Wilson and others, 2003) and increases antibiotic resistance (Sprague and Battaglin, 2005).

Pharmaceuticals

More than 3.9 billion prescription drugs are purchased annually in the United States, and the average American takes more than 12 prescription drugs each year (Kaiser Family Foundation, 2010). Fifty to 90 percent of the active ingredients in these pharmaceuticals passes through the body and is excreted as either the parent compound or its metabolites (Lubliner and others, 2008). From there, these pharmaceuticals enter the wastewater stream, to either a WWTP or a septic system. Besides excretion, the other main path to the environment for pharmaceuticals is disposal. It was once recommended that consumers dispose of pharmaceuticals through either their drain or toilet, but this outdated practice is now discouraged in an effort to reduce the amount of these compounds travelling to the WWTP. Many states are trying to develop drug take-back programs, but federal narcotic regulations complicate the process. In the meantime, consumers are asked to mix unused pharmaceuticals with coffee grounds or kitty litter and dispose of them in the trash.

Residential homes, long-term care facilities, health-care facilities such as hospitals, and veterinary clinics are current sources for pharmaceuticals reaching the environment (Hubbard, 2007), but landfill leachates and garbage-incinerator emissions may be emerging sources as society tries to deal with the disposal issue. Most of these sources use WWTPs as the pathway for reaching the receiving waters. Removal of this class of compounds from the waste stream is complicated by the varying chemical nature of the compounds. Typical treatment techniques used by WWTPs remove some of these pharmaceuticals but are ineffective on others, such as carbamazepine. The amount of pharmaceuticals entering the environment may be reduced by consumers who use fewer pharmaceuticals or select "greener" options (Lubick, 2010), and by proper disposal of pharmaceuticals.

Pharmaceuticals, by intent, are biologically active, therefore, although their exact effects on wildlife are not yet fully documented, their presence in the environment would be expected to have adverse ecological effects (Williams, 2005). Pharmaceuticals and other contaminants delivered through WWTP effluent can be considered to have "pseudo-persistence" because of the continual input of these compounds (Smital, 2008). The effects of continuous low-level exposure to these pharmaceuticals, particularly during sensitive life stages, as well as long-term exposure to the complex mixtures in these effluents are further unknowns (Daughton and Ternes, 1999).

Halogenated Compounds

Polybrominated diphenyl ethers (PBDEs) are man-made chemicals used as flame retardants in electronics, building materials, seat cushions, and clothing. PCBs are stable, nonflammable chemicals used as insulators and cooling compounds in electric equipment and have been used in other products like paint, inks, and pesticides. PCBs and PBDEs have similar structures and are similar toxicologically, causing problems in marine and freshwater fish ranging from neurotoxicity to hormone disruption (Lower Columbia River Estuary Partnership, 2007). A recent study found that exposure to PBDEs was associated with depressed levels of thyroid-stimulating hormone in pregnant women, the health implications of which are unknown (Chevrier and others, 2010). Both PCBs and PBDES are persistent, hydrophobic compounds that do not degrade or dissolve readily in water and tend to bioaccumulate in fatty tissues and have been detected in soil, air, water, sediment, and bodies of fish, wildlife, and people. Johnson and others (2007) measured PCBs in the tissue of juvenile salmon from the lower Columbia River downstream of the industrial and urban Portland/Vancouver area at concentrations exceeding adverse-effects thresholds. Recently, PBDEs have been detected in multiple arctic species (Arctic Monitoring and Assessment Programme and Arctic Council Action Plan to Eliminate Pollution of the Arctic, 2005), illustrating the ubiquitous nature of these halogenated compounds.

Currently Used Pesticides

Currently used pesticides include herbicides and insecticides that often can be found in any home or garage and are used for pest control (flea medicine for pets often contains fipronil), garden care (household insect spray often contains pyrethroids such as permethrin), or general weed maintenance (Casoron® contains dichlobenil and Pendulum® contains pendimethalin). Although pesticides often are discussed as a pollutant of concern in agricultural areas, urban areas can be a source as well because of residential use, commercial-landscape use, and road maintenance.

Organophosphates and carbamates have been shown to have sublethal effects on salmon, causing problems with olfaction, homing, and predator avoidance (Sandahl and others, 2007). Pesticides are rarely detected alone and often occur in the environment in mixtures. Mixtures of pesticides can have an additive or synergistic effect when they are together in the environment (Lower Columbia River Estuary Partnership, 2007). Laetz and others (2009) determined that mixtures of diazinon, chlorpyrifos, malathion, carbaryl, and carbofuran—the most extensively used pesticides in California and the Pacific Northwest—significantly inhibit acetylcholinesterase activity more when they are present together than when they are present individually. This acetylcholinesterase inhibition can interfere with survival behaviors and essential reactions to stimuli; therefore, the presence of these mixtures may be affecting salmon recovery more than expected.

Polycyclic Aromatic Hydrocarbons

PAHs are persistent, widespread organic contaminants that are in petroleum products, creosote-treated wood, paints and dyes, or are created through incomplete combustion (Lower Columbia River Estuary Partnership, 2007). PAHs tend to adsorb to sediments, which can then act as reservoirs for future transport. Benthic invertebrates living in this sediment can bioaccumulate PAHs and pass them on to their predators. In vertebrates, like fish, however, PAHs do not bioaccumulate but are metabolized, and some PAHs are known or suspected carcinogens for vertebrates (Johnson and others, 2002). Parking lots treated with coal-tar-based sealcoat have been shown to be a major source of PAHs (VanMetre and others, 2009); therefore, stormwater passing over these surfaces can transport PAHs to the receiving streams.

Trace Elements and Mercury

Trace elements are metals and other natural chemicals that can be toxic even at low concentrations and aquatic biota have little need for them. For this report, these compounds include arsenic, cadmium, chromium, copper, lead, nickel, selenium, silver, and zinc. Although trace elements are naturally occurring, they also can be introduced through industrial uses and motor vehicles. For instance, copper and zinc are contributed to roads and other impervious services from brake pads, tires, and vehicle exhaust (Davis and others, 2001), and then stormwater runoff transports these deposits to its receiving waters (Sandahl and others, 2007). Copper has been shown to have sublethal effects on salmon behavior through effects on olfaction even at water concentrations as low as 2 µg/L (Baldwin and others, 2003). Another commonly detected trace element, cadmium, bioaccumulates in reproductive organs of fish and disrupts important endocrine processes, especially those involved in synthesis, release and metabolism of hormones (Tilton and others, 2003).

Mercury in the environment is never destroyed but simply cycles between chemical and physical forms. In the aquatic environment, mercury is converted to a more toxic form, methylmercury, which is most often detected in fish. Methylmercury is a known neurotoxin and studies have shown that environmentally realistic concentrations of methylmercury can impair the reproductive cycle in fish (Drevnick and Sandheinrich, 2003).

Contaminant Concentrations in Wastewater-Treatment-Plant Effluent

Anthropogenic Organic Compounds

Anthropogenic organic compounds were measured at each WWTP (table 9), but an obvious pattern based on population (table 1) did not emerge from the results. For instance, it might be expected that a larger number of compounds may be detected at larger population centers with a smaller number of detections occurring at smaller population centers. If this were true, large differences may be expected between the results for Umatilla (roughly 5,000 people) and for Portland (roughly 500,000 people), but this is not often the case. The data indicate that there are many more variables (for instance treatment technology) affecting the presence and concentration patterns than simply population. Flame retardants and steroids were consistently detected in all WWTP-effluent samples, although few pesticides or PAHs were detected, except at Longview. Longview also was notable because it had the greatest number of detections and the concentrations were usually among the highest, particularly for the personal-care-product compounds.

The compounds at all WWTPs (table 9) were two flame retardants [tri(2chloroethyl)] phosphate and tri(dichloroisopropyl)phosphate), a fixative often included in sunscreen (benzophenone), a suspected endocrine disruptor used as a fumigant (1,4-dichlorobenzene), an extensively used musk (galaxolide), two fecal indicators (cholesterol and 3-*beta*-coprostanol), and a plant sterol (*beta*-sitosterol). Compounds also found in all samples except those collected in Umatilla included a known endocrine disrupter and detergent metabolite (NP2EO), another flame retardant (tributyl phosphate), another extensively used musk (tonalide), a common ingredient in cosmetics and pharmaceuticals (triethyl citrate), and a plasticizer (triphenyl phosphate). The effluent from all cities except Richland contained a popular stimulant (caffeine) and a suspected endocrine disruptor and disinfectant used in most hand soaps (triclosan). These examples show the variety and prevalence of these compounds in WWTP effluent.

Effluent stream past ultraviolet disinfection at the City of Hood River Wastewater-Treatment Plant, Oregon, December 2008.

Many of these compounds are somewhat hydrophobic, indicating that they prefer to be associated with sediment or solids rather than the water phase. Observing the amount of these compounds transported in the unfiltered effluent raises questions about what types of compounds may be detected in the biosolids generated by the WWTPs, and at what concentrations. Kinney and others (2006) surveyed biosolids destined for land application from 9 cities across the United States and detected 25 AOCs present. All of these compounds also were found in this study, except fluoxetine, which was not measured, indicating that biosolids may be a source of contamination that can be studied in the future in the Columbia River Basin (Oregon Department of Environmental Quality, 2011).

Table 9. Anthropogenic organic compounds detected in unfiltered wastewater-treatment-plant effluent, Columbia River Basin, Washington and Oregon, 2008–09.

[Station names are shown in table 2. Concentrations reported in micrograms per liter. See table A2 for a listing of anthropogenic organic compounds analyzed and their reporting limits. Present, presence is verified, but concentration is not quantified; sometimes the reporting limit for an individual sample is raised because of matrix interference, these instances of non-detection are shown as less-than (<) the raised reporting limit. **Abbreviations:** ND, not determined because of matrix interferences; E, estimated; –, not detected]

Analyte	Wenatchee (2008)	Wenatchee (2009)	Richland	Umatilla	The Dalles	Hood River	Portland (a.m.)	Portland (noon)	Portland (p.m.)	Vancouver	St. Helens	Longview
Detergent metabolites												
4-Cumylphenol	Present	–	–	–	–	–	–	–	–	–	–	–
para-Nonylphenol (total)	Present	–	Present	–	–	Present	Present	Present	Present	–	E 2.7	Present
4-Nonylphenol monoethoxylate (sum of all isomers) [NP1EO]	< 2.0	–	Present	< 2.0	Present	Present	E 5.8	E 6.3	E 4.4	Present	Present	Present
4-Nonylphenol diethoxylate (sum of all isomers) [NP2EO]	E 2.2	E 1.5	E 2.6	–	E 3.8	E 2.1	E 14	E 13	E 9.3	Present	E 1.3	E 1.7
4-*tert*-Octylphenol	–	Present	–	–	–	–	–	–	–	–	E 0.21	E 0.16
4-*tert*-Octylphenol diethoxylate (OP2EO)	< 0.78	E 0.37	–	–	E 1.0	E 0.32	E 13	E 1.2	E 0.90	E 0.38	E 0.31	E 0.85
4-*tert*-Octylphenol monoethoxylate (OP1EO)	–	Present	–	–	E 0.80	–	E 0.90	E 0.84	E 0.81	–	–	E 0.55
Flame retardants												
Tri(2-butoxyethyl)phosphate	E 0.74	E 0.42	–	–	E 3.4	E 0.54	E 2.4	E 1.9	E 2.3	E 11	E 2.2	E 1.9
Tri(2-chloroethyl)phosphate	0.65	0.44	E 0.15	0.46	0.26	0.45	0.41	0.41	0.3	E 0.16	Present	0.3
Tri(dichloroisopropyl)phosphate	0.64	0.69	E 0.18	0.39	0.43	0.52	0.39	0.45	0.32	E 0.28	E 0.12	0.33
Tributyl phosphate	Present	0.2	Present	–	Present	Present	E 0.60	E 0.38	Present	Present	0.33	0.42
Miscellaneous—antioxidants, solvents, or multiple uses												
Anthraquinone	–	–	–	Present	–	–	–	–	0.22	–	E 0.15	–
Bromoform	–	–	Present	–	Present	–	–	–	–	–	–	–
3-*tert*-Butyl-4-hydroxyanisole (BHA)	–	Present	–	–	–	–	–	–	–	–	–	–
Caffeine	E 0.073	Present	–	Present	E 0.076	0.28	1.6	0.92	1.2	Present	7	20
Cotinine	–	–	–	–	Present	–	Present	Present	Present	–	Present	E 0.34
para-Cresol	–	Present	–	0.25	–	–	–	–	Present	–	Present	E 0.12
Isophorone	–	–	–	–	–	–	–	Present	Present	–	–	–
Isopropylbenzene (cumene)	–	–	–	–	–	–	–	–	–	–	–	E 0.032
d-Limonene	–	–	–	–	–	–	–	–	–	–	–	E 0.078
5-Methyl-1H-benzotriazole	E 0.59	Present	E 0.36	–	E 0.52	E 0.80	E 0.74	E 0.62	E 0.62	E 0.57	–	E 0.89
Pentachlorophenol	E 0.89	Present	–	–	–	–	E 0.79	E 0.77	–	–	Present	Present
Tetrachloroethylene	–	Present	Present	–	Present	–	Present	Present	Present	–	–	Present
Personal care products												
Acetophenone	–	E 0.14	–	–	–	–	–	–	–	–	–	E 0.34
Benzophenone	0.27	0.24	Present	Present	0.26	0.2	0.28	0.26	0.22	E 0.10	E 0.19	0.30
Camphor	–	–	–	–	–	–	–	–	–	–	Present	1.3
1,4-Dichlorobenzene	E 0.26	0.33	Present	E 0.04	E 0.40	E 0.24	E 0.40	E 0.42	E 0.41	E 0.10	Present	0.88
Galaxolide	1.5	2.5	E 1.5	1.3	1.2	1.1	1.0	1.2	0.97	E 1.2	0.38	1.2

Table 9. Anthropogenic organic compounds detected in unfiltered wastewater-treatment-plant effluent, Columbia River Basin, Washington and Oregon, 2008–09.—Continued

[Station names are shown in table 2. Concentrations reported in micrograms per liter. See table A2 for a listing of anthropogenic organic compounds analyzed and their reporting limits. Present, presence is verified, but concentration is not quantified; sometimes the reporting limit for an individual sample is raised because of matrix interference, these instances of non-detection are shown as less-than (<) the raised reporting limit. **Abbreviations:** ND, not determined because of matrix interferences; E, estimated; –, not detected]

Analyte	Wenatchee		Richland	Umatilla	The Dalles	Hood River	Portland			Vancouver	St. Helens	Longview
	(2008)	(2009)					(a.m.)	(noon)	(p.m.)			
Personal care products—Continued												
Indole	–	Present	–	–	–	Present	–	–	–	–	–	E 0.11
Menthol	<0.50	–	–	<0.25	<0.33	<0.24	<0.29	<0.51	–	–	0.32	3.2
3-Methyl-1H-indole (skatol)	Present	Present	–	Present	Present	Present	Present	Present	Present	–	–	Present
Phenol	–	–	–	Present	–	Present	–	Present	–	–	–	Present
Tonalide	Present	E 0.15	E 0.18	–	E 0.11	E 0.12	Present	E 0.14	E 0.11	Present	Present	E 0.11
Triclosan	0.56	0.44	–	E 0.13	0.24	0.25	0.22	0.29	0.21	E 0.29	0.76	0.64
Triethyl citrate (ethyl citrate)	0.51	0.52	E 0.13	0.44	0.44	0.23	0.58	0.59	0.43	E 0.26	E 0.13	0.51
Pesticides												
Atrazine	E 0.17	Present	–	–	–	E 0.077	E 0.090	E 0.14	Present	–	E 0.12	–
3,4-Dichlorophenyl isocyanate	–	–	Present	Present	Present	–	–	–	Present	–	–	–
Metalaxyl	–	–	–	E 0.14	–	–	–	–	–	–	–	–
N,N-diethyl-meta-toluamide (Deet)	E 0.15	Present	–	–	Present	–	0.32	0.31	0.24	–	E 0.15	0.3
Plasticizers												
Bisphenol A	Present	Present	–	–	ND	–	–	–	ND	Present	Present	0.58
Diethyl phthalate (DEP)	–	Present	–	–	–	–	–	0.23	–	–	Present	E 0.19
bis-(2-Ethylhexyl) phthalate (DEHP)	–	–	E 1.2	E 1.7	–	–	–	E 1.3	–	–	Present	Present
Triphenyl phosphate	Present	E 0.10	Present	Present	Present	Present	Present	Present	Present	Present	Present	Present
Polycyclic aromatic hydrocarbons (PAHs)												
Anthracene	–	–	–	–	–	–	–	–	–	–	Present	–
2,6-Dimethylnaphthalene	–	–	–	–	–	–	–	–	–	–	–	Present
Fluoranthene	E 0.11	–	–	–	–	–	–	–	–	–	–	–
1-Methylnaphthalene	–	–	–	–	–	–	–	–	–	–	–	E 0.044
2-Methylnaphthalene	–	–	–	–	–	–	–	–	–	–	–	E 0.047
Naphthalene	–	–	–	–	–	–	–	–	–	–	–	E 0.073
Phenanthrene	–	–	–	–	–	–	–	–	–	–	Present	–
Pyrene	–	–	Present	–	Present	Present	Present	Present	Present	–	–	–
Steroids												
Cholesterol	E 6.3	Present	E 4.5	Present	E 3.1	Present	E 4.1	E 3.8	E 3.4	Present	E 3.6	E 3.4
3-beta-Coprostanol	E 5.8	Present	Present	Present	Present	Present	E 4.5	E 4.3	E 3.6	Present	Present	Present
beta-Sitosterol	E 3.2	E 1.6	E 2.2	E 1.1	E 0.79	E 0.87	E 1.5	E 1.2	E 1.2	E 0.86	E 2.2	E 1.1
beta-Stigmastanol	E 1.1	E 0.55	E 0.29	–	–	–	–	–	–	–	E 0.97	Present

Pharmaceuticals

Pharmaceuticals are expected to be in WWTP effluent because of the amount that passes through the body. In filtered effluent samples, 14 human-health pharmaceuticals were analyzed for and all but albuterol and warfarin were detected from at least one city (table 10). Two pharmaceuticals were detected in samples collected at all WWTPs—carbamazepine, a prescription drug used to treat epilepsy, bipolar disorder, and attention deficit hyperactivity disorder; and diphenhydramine, a common ingredient in over-the-counter medicines used for allergy relief or as a sleep aid, although trimethoprim, a prescription antibiotic, was detected at all WWTPs except Umatilla.

A few studies have determined that the wastewater-treatment process removes or degrades less than 20 percent of the carbamazepine entering the plant (Heberer, 2002; Rounds and others, 2009). Therefore, it is not surprising that carbamazepine was the most frequently detected pharmaceutical in this study, the Chesapeake Bay (Pait and others, 2006), Las Vegas Wash (Boyd and Furlong, 2002), and streams in Germany (Ternes, 1998). Little work has been done to assess toxicity of carbamazepine in the aquatic environment, but one study showed that acute toxicity for bacteria, algae, diatoms, and crustaceans was at the milligrams per liter level (Ferrari and others, 2004), much higher than the concentrations reported here.

Diphenhydramine and trimethoprim have been reported in the Columbia River Basin in trace amounts in both the water column and bed sediments (Morace, 2006; Nilsen and others, 2007). It is difficult to determine potential aquatic effects of diphenhydramine because of its multiple modes of action on histamines, acetylcholine, and transporter receptors (Berninger and others, 2011). When tested on the aquatic invertebrate *Daphnia magma*, Berninger and others (2011) found a no-observed-effect level at the environmentally relevant concentration of 0.8 µg/L, eight times higher than the highest concentration measured in this study. The highest detection frequency of any antibiotics was for trimethoprim in a study of streams in the United States (Kolpin and others, 2002), and was detected in approximately 90 percent of WWTP-effluent samples and 20 percent of stream samples in a German study (Hirsch and others, 1999). Lindberg and others (2007) determined that a trimethoprim concentration of 16 mg/L in water caused growth inhibition in 50 percent of a test species of green algae. Lam and others (2004) calculated trimethoprim to have a half-life of 5.7 days.

Table 10. Pharmaceuticals detected in filtered wastewater-treatment-plant effluent, Columbia River Basin, Washington and Oregon, 2008–09.

[Station names are shown in table 2. Concentrations reported in micrograms per liter. See table A3 for a listing of pharmaceuticals analyzed in filtered-effluent samples and their reporting limits. Present, presence is verified, but concentration is not quantified. **Abbreviations:** E, estimated; –, not detected]

Analyte	Wenatchee (2008)	Wenatchee (2009)	Richland	Umatilla	The Dalles	Hood River	Portland (a.m.)	Portland (noon)	Portland (p.m.)	Vancouver	St. Helens	Longview
Acetaminophen	–	–	–	–	–	–	–	–	–	–	–	E 2.6
Caffeine	–	–	–	–	–	0.22	E 1.3	0.65	2.2	–	E 12	E 30
Carbamazepine	0.1	0.098	0.052	0.12	0.047	0.077	E 0.030	E 0.023	E 0.028	E 0.037	Present	E 0.052
Codeine	–	–	–	–	–	0.042	0.17	0.19	0.13	–	Present	–
Cotinine	–	–	–	–	0.098	–	–	–	–	–	0.071	0 15
Dehydronifedipine	–	–	–	Present	–	–	–	–	–	Present	–	–
Diltiazem	E 0.089	Present	E 0.047	–	E 0.042	E 0.056	–	Present	Present	E 0.043	–	–
1,7-Dimethylxanthine	–	–	–	–	–	–	–	–	–	–	E 0.40	E 5.1
Diphenhydramine	0.090	0.059	E 0.025	Present	0.11	0.082	0.075	0.064	0.056	0.10	E 0.033	E 0.031
Sulfamethoxazole	–	Present	Present	E 0.12	–	–	–	–	–	Present	–	Present
Thiabendazole	0.22	0.57	–	–	–	–	–	–	–	–	–	–
Trimethoprim	0 15	0.19	0.10	–	0.11	0.12	0.089	0.079	0.076	0.073	E 0.020	0.072

Sampling effluent at the Vancouver Westside Wastewater-Treatment Plant, Washington, December 2008.

Estrogenicity

Samples collected from WWTPs in 2008 were screened for total estrogenicity of the compounds in the sample using YES (David Alvarez, U.S. Geological Survey, written commun., November 6, 2009). EEQ values are intended to measure the potential biological effects of the mixture of chemicals present in the sample, and therefore, are not compared to individual compounds in this report.

The compounds that make up these complex mixtures likely have not all been measured in this study. Likewise, the interactions of these compounds in terms of synergistic or antagonistic effects are not defined well enough to analyze relationships between the EEQ and a list of compounds. A large suite of natural and synthetic hormones and phytoestrogens that may be available in the environment could affect the EEQ. Manmade chemicals like phthalates, alkylphenol surfactants, and potentially musks or fragrances also may have estrogenicity that could contribute to the EEQ measure. For all of these reasons, the EEQs measured for this study are not related to compound concentrations, but are used to indicate a measure of the potential for biological effects in the ecosystem.

For this study, the unfiltered WWTP-effluent samples sent to CERC were filtered and the YES was performed on both the filtered effluents and the solids filtered from these samples (table 11). The EEQs measured in the filtered effluents fall into groupings related to each WWTP's discharge rate, with higher EEQs (table 11) associated with plants with higher flow (table 12). Most of the solid samples did not show quantitative levels of estrogenicity. This indicates that the hormonally active compounds are likely dissolved in the effluent water (David Alvarez, U.S. Geological Survey, written commun., November 6, 2009).

Table 11. Estrogenicity in wastewater-treatment-plant effluent samples, instantaneous loadings, and calculated concentrations in the Columbia River, Columbia River Basin, Washington and Oregon, December 2008.

[Station names are shown in table 2. Slight, estrogenic response was observed above the 99 percent confidence limits, but below a measurable threshold. A 7Q10 flow is not available for Umatilla, so 78,000 ft³/s is used for these calculations. **Abbreviations:** EEQ, estimated estradiol equivalents; ng E2/L, nanograms of 17β-estradiol per liter; ng E2/g, nanograms of 17β-estradiol per gram; Mgal/d, million gallons per day; 7Q10, the lowest streamflow for 7 consecutive days that occurs on average every 10 years; ft³/s, cubic foot per second; g/d, gram per day; –, not detected]

Analyte	Wenatchee	Richland	Umatilla	The Dalles	Hood River	Portland (a.m.)	Portland (noon)	Portland (p.m.)	Vancouver
EEQ in filtered water (ng E2/L)	550	760	91	230	55	1,200	1,800	1,400	780
EEQ in solids filtered from water (ng E2/g)	Slight	–	–	–	–	Slight	Slight	Slight	–
Daily plant discharge for sampling date (Mgal/d)	3.1	5.44	0.545	1.7	0.893	49	49	49	10
Columbia River 7Q10 flow (ft³/s)	51,557	52,700	78,000	80,637	74,000	79,436	79,436	79,436	79,436
Instantaneous load (g/d)	6.5	16	0 19	1.5	0.19	223	334	260	30
EEQ in Columbia River (ng E2/L)	0.051	0.12	0.0010	0.0075	0.0010	1.1	1.7	1.3	0.15

Table 12. Physical properties and suspended-sediment results for wastewater-treatment-plant effluent, Columbia River Basin, Washington and Oregon, 2008–09.

[Station names are shown in table 2. **Abbreviations:** Mgal/d, million gallons per day; mg/L, milligrams per liter; µm, micrometer; NA, not analyzed]

City	Date	Time	Daily plant discharge for sampling date (Mgal/d)	Water temperature (degrees Celsius)	pH (standard units)	Suspended sediment (mg/L)	Suspended sediment (percent finer than 63 µm)
Wenatchee	12-02-08	1010	3.1	15.6	6.9	4	98
	12-01-09	0850	2.9	18.0	7.4	3	92
Richland	12-04-08	0900	5.4	18.3	7.2	5	64
	12-02-09	0820	5.8	NA	NA	NA	NA
Umatilla	12-03-08	0840	0.54	16.9	7.4	2	93
	12-02-09	0950	0.53	NA	NA	NA	NA
The Dalles	12-05-08	0830	1.7	15.8	7.0	2	96
	12-02-09	1200	1.5	NA	NA	NA	NA
Hood River	12-10-08	0950	0.89	14.0	6.9	4	95
	12-02-09	1310	0.92	NA	NA	NA	NA
Portland (a.m.)	12-09-08	0900	49	15.6	8.8	3	95
(noon)	12-09-08	1150	49	NA	NA	5	99
(p.m.)	12-09-08	1500	49	NA	NA	4	94
Portland	12-10-09	0840	58	NA	NA	NA	NA
Vancouver	12-08-08	0940	10	18.5	7.4	3	97
	12-02-09	1510	9.7	NA	NA	NA	NA
St Helens	12-03-09	0900	6.9	NA	NA	7	96
Longview	12-08-09	0810	6.9	NA	NA	6	94

The estrogenicity levels measured in this study are well above levels that have been shown to cause effects in aquatic biota. In Swiss midland rivers, brown trout showed a relationship between sites with higher EEQ values and male fish with elevated vitellogenin levels (Vermeirssen and others, 2005). Colman and others (2009) showed that short-term exposure to estrogenic compounds could alter reproductive success in male zebrafish. In their experiment, one-half of the dominant male zebrafish in waters with EEQ levels of 50 ng/L relinquished their paternal dominance. Kidd and others (2007) designed a study in which Canadian experimental lakes were dosed with varying levels of the synthetic estrogen 17α-ethynylestradiol to study the long-term effects on fathead minnows. Chronic exposure to low concentrations (5–6 ng/L) led to the feminization of the males through the production of vitellogenin, and ultimately, the near extinction of this species from the lake.

In this study, the estrogenicity was measured in the effluent itself and not the receiving waters where aquatic biota reside. To compare these results to the studies previously discussed, instantaneous estrogenicity loadings were calculated by multiplying the estrogenicity by the daily plant discharge for the WWTP (table 12) and a conversion factor. To determine the resulting increase in estrogenicity in the Columbia River that could be attributed to this incoming load (table 11), the instantaneous load was divided by the 7Q10 streamflow of the Columbia River at that point of discharge (table 2) and multiplied by a conversion factor. The 7Q10 streamflow is a measure used in National Pollutant Discharge Elimination System (NPDES) permit writing that is equal to the lowest streamflow for 7 consecutive days that would be expected to occur once in 10 years. The 7Q10 streamflow is used to determine mixing zones by providing a measure of the streamflow available to dilute any inputs at that point in the river. The resulting estrogenicity in the Columbia River near the Portland WWTP was calculated to be greater than 1 ng/L (table 11), a concentration that may potentially cause endocrine disruption in different aquatic species (Nelson and others, 2007).

Halogenated Compounds

Halogenated compounds were analyzed (table 13) from the solids collected by filtering roughly 10–20 L of effluent from each WWTP. Concentrations of PBDEs were detected at every WWTP, and the highest concentrations detected were for congeners PBDE-47, PBDE-99, and PBDE-100. Most of these values are reported in table 13 as "Present" because of detectable concentrations in the set blanks. The few concentrations reported are much higher than those for any of the other congeners, which is a typical pattern observed in environmental data (Yoqui and Sericano, 2009) because these congeners are the most stable. All of the PBDE congeners detected in this study also have been detected in Osprey eggs collected in Oregon and Washington (Henny and others, 2009).

The highest PBDE concentrations were in Richland and Portland. The Portland PBDE values showed the lowest concentrations occurring in the morning and then the concentrations detected later in the day amounting to roughly two to four times the morning concentrations. This pattern of lower concentrations in the morning and higher concentrations later in the day also was observed in the other halogenated compounds measured. Interestingly, the suspended-sediment concentrations remained fairly low and constant throughout the day. Therefore, this pattern of higher concentrations later in the day is not due to changes in the amount of suspended sediment, but rather the amount of associated with these solids. The only WWTP where PCBs were detected was in Wenatchee. The reason for this is not known, but, since June 2007, there have been fish advisories for mountain whitefish in the lower Wenatchee River based on their elevated PCB concentrations (Washington State Department of Health, 2007). Several pesticide compounds were detected in these solids. These types of pesticides are more hydrophobic, so these compounds likely would be detected in these samples and in biosolids, which were not analyzed in this study.

Currently Used Pesticides

Few currently used pesticides were detected in WWTP-effluent samples (table 14). The primary compounds detected were fipronil and its degradates, which were detected in all WWTPs except Wenatchee. Fipronil is an insecticide used to control common household pests like ants, beetles, cockroaches, and other insects, and can be included in topical pet-care products used to control fleas (National Pesticide Information Center, 2009). A common herbicide degradate,

3,4-dichloroaniline, was detected at all WWTPs, although propiconazole, the only fungicide detected in this study (only at St. Helens), but at concentrations approaching 10 μg/L (sum of cis- and trans-propiconazole). Propiconazole is a wood preservative designed to prevent fungal decay in above ground applications (U.S. Environmental Protection Agency, 2008), and the St. Helens WWTP gets a large amount of wastewater from the nearby Boise Cascade pulp and paper mill.

Mercury

Because of SB 737, mercury and methylmercury were added to the analyte list for WWTP-effluent samples collected in November 2009 (table 15). The highest total mercury concentrations were measured at The Dalles and Vancouver. Both concentrations were greater than 12 ng/L, the chronic criterion for freshwater aquatic life (Washington State Department of Ecology, 2003; Oregon Department of Environmental Quality, 2010b). This chronic criterion is the average 4-day concentration, whereas the acute criterion (1-hour average concentration) of 2,400 ng/L was not exceeded in this study. Methylmercury, the bioavailable form of mercury in the environment, concentrations were all fairly low (0.40 ng/L or less); the highest concentration was detected at The Dalles.

Synopsis

Of the 210 compounds analyzed in the WWTP-effluent samples, 112 or 53 percent were detected in at least 1 sample (fig. 3). Most compound classes had a greater than 80-percent detection rate, emphasizing the complex mixtures of contaminants present in WWTP effluent. Interesting patterns emerge when these percentages of detection are displayed by individual WWTP (table 16). Although there are variations in the individual composition of the samples for each plant, there are many similarities in the frequency of detections across the plants. For example, the detection frequency for flame retardants at all plants was 65–82 percent. Similarly, personal care products, pesticides, steroids, pharmaceuticals, and miscellaneous compounds showed similar detection frequencies amongst the plants. These similarities illustrate that although there are differences between the plants based on location, population, treatment type, and plant size, many of the results are similar. Of notable difference were the PAHs, which were more sporadic, and the PCBs, which were detected only at Wenatchee, St. Helens, and Longview.

Table 13. Halogenated compounds detected in solids filtered from wastewater-treatment-plant effluent, Columbia River Basin, Washington and Oregon, 2008–09.

[Station names are shown in table 2. Concentrations reported in nanograms per liter. See table A1 for a listing of halogenated compounds analyzed and their reporting limits. Present, presence is verified, but concentration is not quantified; sometimes the reporting limit for an individual sample is raised because of matrix interference, these instances of non-detection are shown as less than (<) the raised reporting limit. **Abbreviations:** L, liter; –, not detected; ND, not determined because of poor compound recoveries]

Analyte	Wenatchee	Richland	Umatilla	The Dalles	Hood River	Portland (a.m.)	Portland (noon)	Portland (p.m.)	Vancouver	St. Helens	Longview
Volume filtered (L)	21	19	18	18	19	20	10	9	18	12	20
Polybrominated diphenyl ethers (PBDEs) or brominated flame retardants											
Dechlorane plus	Present	Present	–	–	–	–	–	–	–	Present	–
Firemaster 680	0.02	0.07	0.02	0.04	0.13	0.05	0.03	0.08	0.06	0.05	–
PBDE-47	Present	Present	Present	Present	Present	Present	Present	Present	Present	Present	Present
PBDE-66	0.07	0.22	0.14	0.15	0.09	0.10	0.02	0.11	0.45	0.04	0.05
PBDE-85	0.07	0.28	0.09	0.09	0.08	0.17	0.09	0.21	0.21	0.08	0.12
PBDE-99	Present	Present	Present	Present	Present	Present	Present	4.8	Present	Present	Present
PBDE-100	Present	1.6	Present	Present	Present	Present	Present	Present	Present	Present	Present
PBDE-138	0.02	0.06	0.04	0.03	0.03	0.05	0.02	0.08	0.06	0.03	–
PBDE-153	0.12	0.40	0.25	0.14	0.15	0.06	0.11	0.36	0.27	0.16	0.16
PBDE-154	0.12	0.38	0.27	0.15	0.15	0.25	0.08	0.31	0.28	0.15	0.16
PBDE-183	–	0.05	0.03	0.03	0.04	0.04	0.02	0.04	0.04	–	–
Polychlorinated biphenyls (PCBs)											
PCB-101	–	–	–	–	–	–	–	–	–	–	Present
PCB-146	0.01	–	–	–	–	–	–	–	–	–	–
PCB-170	0.01	–	–	–	–	–	–	–	–	–	–
PCB-174	0.01	–	–	–	–	–	–	–	–	–	–
PCB-177	0.01	–	–	–	–	–	–	–	–	–	–
PCB-180	0.02	–	–	–	–	–	–	–	–	0.02	0.01
PCB-183	Present	–	–	–	–	–	–	–	–	–	–
PCB-187	0.01	–	–	–	–	–	–	–	–	–	–
PCB-194	Present	–	–	–	–	–	–	–	–	–	–
Herbicides and insecticides											
cis-Chlordane	0.03	0.19	0.02	0.08	0.05	0.10	–	0.05	0.05	0.07	0.09
trans-Chlordane	0.02	0.19	0.01	0.05	0.03	0.08	0.02	0.06	0.03	0.01	0.05
Chlorpyrifos	–	–	–	–	–	0.18	0.18	0.43	–	0.04	0.03
Cyfluthrin	–	0.26	–	0.18	–	0.26	0.07	0.41	–	–	–
lambda-Cyhalothrin	0.02	Present	–	–	–	Present	–	Present	–	–	–
Desulfinylfipronil	–	Present	Present	0.02	0.02	Present	–	Present	0.07	–	–
Dieldrin	0.01	0.17	–	–	0.05	0.14	–	0.09	0.08	–	< 0.04
alpha-Endosulfan	0.01	–	–	–	–	–	–	–	–	–	–
Fipronil	ND	0.22	0.06	0.17	0.20	0.99	0.35	0.77	1.4	ND	0.05
Fipronil Sulfide	–	0.03	0.02	Present	0.02	0.06	0.02	0.06	0.08	0.04	0.01
cis-Nonachlor	Present	–	–	–	–	–	–	–	–	–	0.01
trans-Nonachlor	0.01	0.10	Present	0.03	–	0.04	–	0.04	–	0.01	0.03
Pentachloroanisole	Present	Present	–	–	Present	–	–	–	0.85	–	Present
Trifluralin	Present	–	–	–	–	0.02	–	0.02	–	–	–
Other compounds											
Hexachlorobenzene (HCB)	–	–	Present	–	–	–	–	–	–	–	–
Methoxy triclosan	1.2	3.7	–	–	4.6	1.7	–	–	13	Present	1.1
Triclosan	Present	Present	–	Present	Present	55	Present	57	86	–	ND

Table 14. Currently used pesticides and degradates detected in filtered wastewater-treatment-plant effluent, Columbia River Basin, Washington and Oregon, December 2009.

[Station names are shown in table 2. Concentrations reported in micrograms per liter. See table A4 for a listing of pesticides analyzed and their reporting limits. Present, presence is verified, but concentration is not quantified; sometimes the reporting limit for an individual sample is raised because of matrix interference, these instances of non-detection are shown as less than (<) the raised reporting limit. **Abbreviations:** E, estimated; –, not detected]

Analyte	Wenachee	Richland	Umatilla	The Dalles	Hood River	Portland	Vancouver	St. Helens	Longview
Fungicides									
cis-Propiconazole	–	–	–	–	–	–	–	E 2.8	–
trans-Propiconazole	–	–	–	–	–	–	–	E 6.9	–
Herbicides and degradates									
Atrazine	–	Present	–	–	–	<0.0095	Present	–	–
3,4-Dichloroaniline[1]	E 0.034	E 0.056	E 0.052	E 0.12	E 0.021	<0.065	E 0.28	E 0.034	E 0.13
Metolachlor	–	–	–	–	–	–	–	E 0.012	–
Prometon	–	Present	Present	–	–	–	–	–	–
Simazine	–	–	–	E 0.005	–	–	–	–	–
Insecticides and degradates									
Carbaryl	–	–	–	E 0.092	–	–	–	–	Present
Desulfinylfipronil[1]	–	Present	Present	Present	Present	–	Present	Present	Present
Fipronil	–	E 0.027	Present	E 0.042	E 0.056	E 0.078	E 0.087	<0.047	E 0.13
Fipronil sulfide[1]	–	Present	Present	Present	Present	Present	Present	0.022	E 0.006
Fipronil sulfone[1]	–	Present	–	–	Present	–	Present	–	0.024
Other compounds									
1-Naphthol[1]	–	–	–	Present	–	–	–	–	Present

[1]Degradate.

Ducks swimming in the clarifier at the City of Wenatchee Wastewater-Treatment Plant, Washington, December 2009.

Table 15. Mercury species in unfiltered wastewater-treatment-plant effluent and stormwater runoff, Columbia River Basin, Washington and Oregon, 2009–10.

[Station names are shown in tables 2 and 3. Outfalls sampled prior to December 2009 were not analyzed for mercury species. Concentrations are reported in nanograms per liter (ng/L). **Symbol:** –, not detected at a reporting limit of 0.04 ng/L]

City or short name	Date	Time	Methyl-mercury	Total mercury
Wastewater-treatment-plant effluent				
Wenatchee	12-01-09	0850	–	4.1
Richland	12-02-09	0820	0.19	4.2
Umatilla	12-02-09	0950	–	1.9
The Dalles	12-02-09	1200	0.40	16
Hood River	12-02-09	1310	–	2.7
Portland	12-10-09	0840	0.14	7.6
Vancouver	12-02-09	1510	0.06	13
St Helens	12-03-09	0900	0.15	3.9
Longview	12-08-09	0810	0.22	10
Stormwater runoff				
Wenatchee	12-21-09	1340	0.09	3.4
Vancouver1	12-16-09	1340	–	8.8
Vancouver2	12-16-09	1210	0.07	15
Willamette1	06-04-10	0840	–	6.3
Willamette2–Dec	12-15-09	1330	0.39	230
Willamette2–May	05-26-10	1410	–	74
Willamette3	12-15-09	1310	–	17
Willamette4	05-26-10	1310	0.11	12
St. Helens	03-30-10	1310	0.07	3.1
Longview	03-30-10	1410	–	2.1

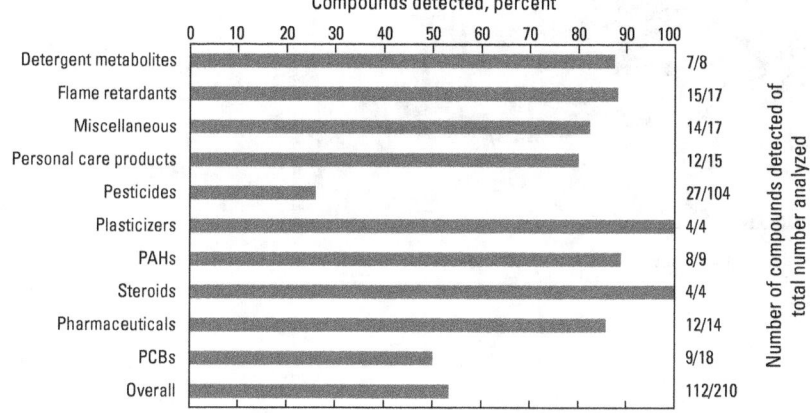

Figure 3. Percentage of compounds detected in wastewater-treatment-plant effluent, Columbia River Basin, Washington and Oregon, 2008–09.

Table 16.   Percentage of compounds detected in each wastewater-treatment plant sampled, Columbia River Basin, Washington and Oregon, 2008–09.

[Station names are shown in table 2]

Compound class	Total number analyzed	Wenatchee	Richland	Umatilla	The Dalles	Hood River	Portland (a.m.)	Portland (noon)	Portland (p.m.)	Vancouver	St. Helens	Longview
Detergent metabolites	8	50	38	0	50	50	63	63	63	38	63	63
Flame retardants	17	82	76	76	82	82	82	82	82	82	82	65
Miscellaneous	17	47	24	29	35	24	35	35	47	24	35	53
Personal care products	15	60	33	47	47	53	47	53	47	40	53	80
Pesticides	104	12	12	18	15	13	9	13	9	16	13	15
Plasticizers	4	100	50	25	50	25	25	75	50	50	100	100
Polycyclic aromatic hydrocarbons (PAHs)	9	0	11	0	11	0	11	11	11	0	22	44
Steroids	4	100	100	75	75	75	75	75	75	75	100	100
Pharmaceuticals	14	43	29	36	36	43	43	36	43	43	50	57
Polychlorinated biphenyls (PCBs)	18	44	0	0	0	0	0	0	0	0	6	11
Overall	210	37	25	28	33	29	29	32	30	30	33	40

Access point for sampling effluent at Vancouver Westside Wastewater-Treatment Plant, Washington, December 2008.

Contaminant Concentrations in Stormwater Runoff

Halogenated Compounds

When compared to the WWTP samples, the stormwater samples analyzed for halogenated compounds had many more performance or matrix issues, resulting in many raised detection limits (table 17). As a result, the patterns observed were quite different from those in the WWTP-effluent samples; these differences may reflect differences between the stormwater runoff and the WWTP-effluent samples. Because of this reporting variability, it is difficult to do a complete analysis of these results. For all classes of compounds in this analysis, however, a few samples stood out. A suspended-sediment sample with a concentration of 834 mg/L was collected during the first rain of autumn 2009 from Umatilla, indicating that a large amount of material was available for compounds to attach (table 18). The median suspended-sediment concentration for all stormwater-runoff samples was 21 mg/L.

The elevated suspended-sediment concentration of 47 mg/L measured in the Willamette2-Dec sample does not provide a complete explanation as to why the highest contaminant concentrations were detected in December and May. The sources contributing to this drainage area may explain the anomalously high concentrations of PBDEs. Of the 51 halogenated compounds detected when all the stormwater sites are considered, 46 were in the Willamette2-Dec sample (table 17). The PBDE concentrations at Willamette2 were roughly twice those detected in the Umatilla sample and the PCB concentrations at Willamette2 were 20–300 times greater than PCB concentrations at any other stormwater locations. The Willamette 2, 3, and 4 stormwater locations are all within the Portland Harbor Superfund area (U.S. Environmental Protection Agency, 2010), but are not in areas that are currently being studied. The Portland Harbor Superfund site is a 10-mile stretch of historical industrial usage in the lower Willamette River where heavy metals, PCBs, PAHs, dioxin/furans, and pesticides have been detected. Investigation and cleanup have been occurring at the site since 1997. Concentrations measured at the Willamette2 site indicate that there are still active inputs of contaminants in the area.

Stormwater outfall at Klindt Point in The Dalles, Oregon, February 2009.

Table 17. Halogenated compounds detected in solids filtered from stormwater runoff, Columbia River Basin, Washington and Oregon, 2009–10.

[Station names are shown in table 3. Outfalls shown in downstream order; concentrations reported in nanograms per liter. See table A1 for a listing of halogenated compounds analyzed and their reporting limits. Present, presence is verified, but concentration is not quantified. **Abbreviations:** L, liter; —, not detected; ND, not determined because of poor compound recoveries; sometimes the reporting limit for an individual sample was raised because of matrix interference, these instances of non-detection are shown as less-than (<) the raised reporting limit]

Analyte	Wenatchee	Richland	Umatilla	The Dalles	Hood River	Portland1	Vancouver1	Vancouver2	Portland2	Willamette1	Willamette2 -Dec	Willamette2 -May	Willamette3	Willamette4	St. Helens	Longview
Volume filtered (L)	2.7	9.4	3.0	10.2	11.3	3.7	5.1	5.5	3.6	4.7	2.9	2.3	5.4	2.3	3.4	4.3
Polybrominated diphenyl ethers (PBDEs) or brominated flame retardants																
Dechlorane Plus	—	—	Present	Present	—	Present	Present	Present	Present	—	5.2	E 1.6	—	<0.64	—	—
Firemaster 680	—	—	0.36`	Present	—	—	—	—	—	—	4.7	0.78	—	0.52	—	—
Pentabromotoluene	—	—	—	—	—	—	—	—	—	—	0.13	—	—	—	—	—
PBDE-47	<11	ND	Present	ND	ND	—	<1.1	<1.2	<0.99	<0.69	18	Present	<0.62	<1.1	<1.1	<0.16
PBDE-66	—	ND	0.45	—	—	—	—	Present	0.11	—	0.43	0.27	—	—	—	—
PBDE-85	—	—	0.60	ND	—	—	—	—	—	—	1.1	0.57	—	—	—	—
PBDE-99	<0.39	ND	9.3	ND	ND	<0.21	<0.63	<0.68	<0.79	Present	22	12	<0.30	Present	<0.67	<0.10
PBDE-100	<0.10	ND	2.0	ND	ND	<0.05	<0.16	<0.18	<0.17	<0.09	4.2	2.3	<0.08	<0.18	Present	—
PBDE-138	—	—	0.14	—	—	—	—	—	—	—	<0.23	0.19	—	—	—	—
PBDE-153	—	—	<0.96	—	—	—	0.06	—	0.11	0.03	2.6	1.1	0.04	0.10	0.08	—
PBDE-154	—	—	0.85	—	ND	—	—	—	0.12	0.03	1.8	0.96	—	—	0.07	—
PBDE-183	—	—	—	—	—	—	—	—	—	—	3.1	0.26	—	<0.16	—	—
Polychlorinated biphenyls (PCBs)																
PCB-52	—	—	—	—	—	—	—	—	—	—	40	—	—	—	—	—
PCB-70	—	—	—	—	—	—	—	—	—	—	32	—	—	—	—	—
PCB-101	—	—	—	—	—	—	—	—	—	—	80	—	—	—	—	—
PCB-110	—	—	—	—	—	—	—	—	—	—	85	—	—	—	—	—
PCB-118	—	—	—	Present	—	—	—	Present	—	—	65	0.54	—	0.50	—	—
PCB-138	—	—	—	—	—	—	—	—	—	—	26	0.42	—	0.31	—	—
PCB-146	—	—	—	Present	—	—	—	—	—	—	8.5	0.16	—	0.08	0.06	—
PCB-149	—	—	—	—	—	—	—	—	—	—	49	<1.1	—	<0.49	—	—
PCB-151	—	—	—	—	—	—	—	—	—	—	8.9	0.20	—	0.08	—	—
PCB-170	—	—	—	Present	—	—	—	—	0.05	—	9.7	0.34	0.05	0.15	—	—
PCB-174	—	—	—	—	—	—	—	—	0.15	—	6.5	0.42	0.05	0.18	—	—
PCB-177	—	—	—	—	—	—	—	—	—	—	4.0	0.26	—	0.15	—	—
PCB-180	—	—	<0.06	Present	—	—	0.06	0.08	—	0.06	14	0.77	0.08	0.23	0.07	0.02
PCB-183	—	—	—	—	—	—	—	—	—	—	3.7	0.18	—	—	—	—
PCB-187	—	—	—	Present	—	—	—	—	0.03	—	5.8	0.37	0.04	0.11	—	—
PCB-194	—	—	<0.95	Present	—	—	—	—	Present	—	1.7	Present	Present	—	—	—
PCB-206	—	—	—	—	—	—	—	—	—	—	0.90	0.12	0.02	0.05	—	—

Table 17. Halogenated compounds detected in solids filtered from stormwater runoff, Columbia River Basin, Washington and Oregon, 2009–10.—Continued

[Station names are shown in table 3. Outfalls shown in downstream order; concentrations reported in nanograms per liter. See table A1 for a listing of halogenated compounds analyzed and their reporting limits. Present, presence is verified, but concentration is not quantified. **Abbreviations:** L, liter; –, not detected; ND, not determined because of poor compound recoveries; sometimes the reporting limit for an individual sample was raised because of matrix interference, these instances of non-detection are shown as less-than (<) the raised reporting limit]

Analyte	Wenatchee	Richland	Umatilla	The Dalles	Hood River	Portland1	Vancouver1	Vancouver2	Portland2	Willamette1	Willamette2		Willamette3	Willamette4	St. Helens	Longview
											–Dec	–May				
Volume filtered (L)	2.7	9.4	3.0	10.2	11.3	3.7	5.1	5.5	3.6	4.7	2.9	2.3	5.4	2.3	3.4	4.3
							Herbicides and insecticides									
cis-Chlordane	–	–	1.3	–	–	–	0.38	–	0.07	0.13	6.1	1.5	–	–	<0.76	–
trans-Chlordane	–	–	0.45	–	–	–	0.23	–	Present	0.04	4.6	0.94	–	–	Present	–
Chlorpyrifos	–	0.16	1.5	–	Present	–	0.93	0.93	0.58	0.31	24	4.0	0.07	0.31	0.12	–
Cyfluthrin	–	3.1	<2.4	–	–	–	<2.2	–	–	–	5.9	3.0	–	–	–	–
lambda-Cyhalothrin	–	–	0.81	–	–	–	–	–	–	–	1.2	0.49	–	–	–	–
Dacthal (DCPA)	–	0.56	7.3	–	–	–	–	–	–	0.03	0.50	0.37	–	Present	–	–
p,p'-DDE	–	–	–	–	–	–	–	–	–	–	6.1	–	–	–	–	–
p,p'-DDT	–	–	–	–	–	–	–	–	–	–	–	17	–	–	–	–
Dieldrin	–	–	–	–	–	–	–	–	–	–	3.0	5.7	–	–	–	–
alpha-Endosulfan	0.10	–	3.7	–	–	–	0.29	0.11	–	0.13	–	0.16	–	Present	–	–
Fipronil	ND	–	ND	–	ND	ND	ND	ND	ND	–	0.92	1.0	ND	–	–	–
Fipronil sulfide	–	–	–	–	–	–	–	–	–	–	–	0.03	–	–	–	–
cis-Nonachlor	–	–	0.38	–	–	–	0.13	0.02	–	–	1.0	0.25	–	–	–	–
trans-Nonachlor	–	–	0.82	–	–	–	0.22	0.03	Present	0.03	2.8	0.72	–	–	–	–
Oxyfluorfen	–	–	3.3	–	–	–	Present	Present	Present	–	–	Present	–	–	–	–
Pendimethalin	–	–	<10	–	–	–	82	150	<0.72	1.45	13	–	–	–	–	–
Pentachloroanisole	–	–	Present	–	–	–	0.57	Present	Present	<0.12	1.2	<0.35	Present	<0.17	Present	<0.04
Pentachloronitro-benzene	–	–	4.1	–	–	–	0.46	0.76	–	–	0.54	<1.2	–	–	0.10	–
Trifluralin	Present	E 0.02	0.58	Present	–	–	1.6	0.16	<0.10	0.17	5.5	7.7	0.04	Present	0.07	–
							Other compounds									
Hexachlorobenzene (HCB)	–	–	Present	–	–	–	Present	Present	–	–	Present	<0.25	<0.03	–	<0.18	–
Tetradifon	–	–	–	ND	ND	ND	–	–	–	–	0.40	–	–	–	–	–
Triclosan	–	Present	ND	ND	ND	ND	–	–	ND	ND	Present	ND	–	ND	<3.9	–

Table 18.   Physical properties and suspended-sediment results for stormwater runoff, Columbia River Basin, Washington and Oregon, 2009–10.

[Station names are shown in table 3. **Abbreviations:** L/min, liters per minute; mg/L, milligrams per liter; μm, micrometer; μS/cm, microSiemens per centimeter; –, not determinded at time of sampling or not analyzed; E, estimated]

Analyte	Estimated discharge from pipe during sampling (L/min)	Suspended sediment (mg/L)	Suspended sediment (percent finer than 63 μm)	pH (standard units)	Specific conductance (μS/cm)
Wenatchee	0.5	5	78	7.2	144
Richland	–	9	82	–	–
Umatilla	–	834	86	–	–
The Dalles	1.8	22	99	–	–
Hood River	1.7	2	59	–	–
Portland1	10	20	95	–	–
Vancouver1	12	61	80	7.3	E 12
Vancouver2	0.8	62	94	7.5	E 38
Portland2	12	10	88	7.6	E 36
Willamette1	–	12	93	7.9	140
Willamette2–Dec	3.3	47	97	6.5	131
Willamette2–May	3.3	36	98	6.8	263
Willamette3	–	162	95	7.1	E 98
Willamette4	2.5	42	98	7.4	112
St Helens	40	5	94	7.1	144
Longview	60	53	99	6.9	243

Herbicides and insecticide detection patterns for the solids filtered from stormwater runoff also follow a pattern of high contaminant concentrations in those samples with high suspended-sediment concentrations—particularly from Umatilla, Vancouver, and Willamette2. The pesticides analyzed on these solids are expected to be associated with and therefore, transported, with sediment and solids. Higher suspended-sediment concentration does not account for the complete pattern observed, however. The second largest suspended-sediment concentration measured in this study was at the Willamette3 site, but only 3 pesticides were detected in this sample compared with the 15 pesticides detected in the Willamette2-Dec sample. This indicates that land use in the drainage area likely is an important factor when examining occurrence and distribution. Examining land-use patterns was outside the scope of this report, but toxic-reduction efforts will be more effective when contaminant occurrence and distribution data is coupled with land-use information from the various stormwater catchments that drain to the Columbia River.

The sample collected at the Willamette2 site in May included p,p'-DDT, which is somewhat unusual because it is not often the parent compound that is detected but rather one of the degradates, often p,p'-DDE. The pipe at the Willamette2 location drains an area that was historically used by a pesticide manufacturer. Therefore, remnants of contaminants from historical land uses may still be contributing to the receiving waters periodically through stormwater runoff. The stormwater samples from both Vancouver locations are notable for one compound, pendimethalin, a pre-emergent herbicide used for control of grassy weeds (Koski, 2008).

Currently Used Pesticides

No single pesticide was detected at all of the locations where filtered-stormwater-runoff samples were analyzed, but among those detected, herbicides and insecticides were detected most often (table 19). This may have been caused by the timing of the samplings; however, no obvious pattern of detections was noted between the autumn/winter and spring storms. The highest numbers of detections were in the Portland/Willamette area, with the most detections and highest concentrations at the Willamette2 site. Detections for several pesticides and PCBs from the Willamette2 site in December and May exceeded chronic freshwater-quality criteria (table 20). Although many of these concentrations are low (less than 1 μg/L), mixtures of some of these pesticides have synergistic and additive effects on salmon health when they occur together (Scholz and others, 2006; Laetz and others, 2009).

Table 19. Currently used pesticides and degradates detected in filtered stormwater runoff, Columbia River Basin, Washington and Oregon, 2009–10.

[Station names are shown in table 3. Concentrations reported in micrograms per liter. See table A4 for a listing of pesticides analyzed and their reporting limits. Present, presence is verified, but concentration is not quantified. **Abbreviations:** E, estimated; –, not detected; sometimes the reporting limit for an individual sample was raised because of matrix interference, these instances of non-detection are shown as less than (<) the raised reporting limit]

Analyte	Wenatchee	Richland	Umatilla	The Dalles	Hood River	Portland1	Vancouver1	Vancouver2	Portland2	Willamette1	Willamette2 –Dec	Willamette2 –May	Willamette3	Willamette4	St. Helens	Longview
Fungicides																
Metalaxyl	–	–	<0.013	–	–	<0.042	<0.014	<0.021	–	<0.0149	<0.020	–	–	–	–	–
Myclobutanil	0.097	–	–	–	–	–	–	–	<0.040	–	–	–	E 0.017	–	–	–
cis-Propiconazole	E 0.059	–	–	–	–	–	<0.019	<0.023	–	E 0.024	<0.056	E 0.055	<0.024	E 0.015	–	E 0.030
trans-Propiconazole	E 0.069	–	–	E 0.016	–	–	<0.022	–	–	E 0.038	<0.053	E 0.069	–	E 0.018	–	E 0.043
Tebuconazole	0.076	–	–	–	–	–	–	–	–	<0.0362	<0.064	0.081	–	–	–	–
Herbicides and degradates																
Atrazine	–	0.022	–	–	–	0.038	<0.009	<0.008	–	0.011	<0.016	0.032	<0.008	–	–	–
4-Chloro-2-methylphenol[1]	–	<0.007	<0.004	–	–	–	–	–	–	–	E 0.322	E 1.1	E 0.008	–	–	–
DCPA (Dacthal)	Present	0.032	0.027	Present	Present	Present	–	–	–	Present	Present	Present	–	Present	–	–
3,4-Dichloroaniline[1]	–	–	E 0.010	E 0.073	–	<0.008	–	–	–	–	–	–	–	–	–	–
EPTC	–	0.029	E 0.005	–	–	–	–	<0.003	–	–	–	–	–	–	–	–
Hexazinone	–	–	–	–	–	–	–	–	0.021	–	–	–	–	–	–	0.021
Metolachlor	–	0.026	E 0.013	–	–	E 0.024	–	–	–	E 0.011	–	–	–	E 0.009	–	–
Pendimethalin	–	–	0.027	–	–	E 0.028	0.10	0.16	–	0.032	<0.043	–	–	–	–	–
Simazine	<0.035	–	–	0.10	–	0.063	<0.008	0.14	0.032	0.051	0.058	0.046	<0.008	0.036	0.054	0.025
Tebuthiuron	–	–	–	0.22	–	<0.042	–	–	–	Present	0.17	0.62	0.056	0.12	–	0.11
Trifluralin	–	–	–	–	–	Present	Present	–	–	Present	E 0.012	E 0.013	–	–	–	–
Insecticides and degradates																
Carbaryl	E 0.13	E 0.20	Present	–	–	E 0.038	Present	–	–	–	E 0.37	E 0.38	–	Present	–	Present
Chlorpyrifos	–	–	<0.018	–	–	E 0.062	–	<0.016	<0.016	–	E 0.017	–	–	–	–	–
Desulfinylfipronil[1]	–	–	–	–	–	Present	Present	Present	–	Present	Present	Present	–	–	–	–
Diazinon	0.13	<0.010	<0.018	–	E 0.007	<0.024	<0.005	–	–	<0.007	0.072	0.78	–	–	–	–
Fipronil	–	–	Present	–	Present	–	Present	–	–	–	E 0.014	E 0.049	–	–	–	–
Malathion	–	–	–	–	–	–	–	–	–	–	0.50	1.3	–	–	–	–
Other compounds																
Ethoprop	–	–	<0.083	–	–	2.9	–	<0.019	<0.019	<0.020	–	–	–	–	–	–
1-Naphthol[1]	–	E 0.026	Present	–	–	–	Present	E 0.016	Present	–	E 0.030	E 0.056	–	–	–	–

[1] Degradate.

Table 20.   Concentrations exceeding freshwater-quality criteria for pesticides and polychlorinated biphenyls (PCBs) in stormwater runoff from the Willamette2 site, Columbia River Basin, Washington and Oregon, December 2009 and May 2010.

[Concentrations are in micrograms per liter. **Acute:** Criteria refer to an instantaneous concentration not to be exceeded at any time. **Chronic:** Criteria refer to a 24-hour average concentration not to be exceeded. References for criteria are: U.S. Environmental Protection Agency (EPA), 2009b; Oregon (OR) Department of Environmental Quality, 2010b; Washington (WA) State Department of Ecology, 2003. **Abbreviations:** CMC, criterion maximum concentration; CCC, criterion continuous concentration; –, no acute criteria]

Analyte	Agency standard	Acute (CMC)	Chronic (CCC)	Concentration range detected in this study	Crtierion exceeded	December	May
Chlordane	EPA, OR, WA	2.4	0.004	0.00007–0.011	Chronic	X	
DDT (and metabolites)	EPA, OR, WA	1.1	0.001	0.0061–0.017	Chronic	X	X
Diazinon	EPA	–	0.17	0.072–0.78	Chronic		X
Dieldrin	WA	2.5	0.002	0.00029–0.0057	Chronic	X	X
Malathion	EPA, OR	–	0.1	0.5–1.3	Chronic	X	X
PCBs	EPA, OR, WA	2	0.014	0.00002–0.44	Chronic	X	

Stormwater drain, Portland, Oregon, October 2009.

Stormwater outfall under west end of St. John's railroad bridge (Willamette2), Portland, Oregon, October 2009.

Polycyclic Aromatic Hydrocarbons

Most concentrations of PAHs in stormwater runoff were low (less than the MDL), and the numbers of detections were consistent among locations, except for Hood River and St. Helens, where no PAHs were detected (table 21). Samples collected in the Portland/Vancouver area had the highest number of detections, likely due to the higher density of potential sources in the area, including industry and automobiles. In contrast, Umatilla is a small town with little urban development; however, the stormwater-runoff sample from Umatilla had a large number of detections. This is likely attributable to the large suspended-sediment concentration (834 mg/L, table 18) and the effects of roadway runoff adjacent to the site.

Trace Elements and Mercury

The 10 trace elements measured in filtered and unfiltered stormwater runoff in this study were detected fairly consistently through all samples (table 22), except for mercury, silver, and selenium, which were detected in only about one-half of the unfiltered-water samples. The detection limits are lower for filtered-water analyses; thus, selenium was detected at low concentrations in most of these samples, but mercury and silver were detected only in a few samples. The ratio of the concentration of each element in the filtered-water sample to that in the unfiltered-water sample indicates that selenium, cadmium, nickel, zinc, arsenic, and copper are transported more readily in the dissolved phase, whereas chromium, silver, lead, and mercury are more often transported in the solid phase (fig. 4). When suspended-sediment concentrations are higher, such as in the samples from Umatilla, Willamette3, and Vancouver1 (table 18), higher concentrations are measured in both phases and the ratio shifts.

Chronic and sometimes acute freshwater-quality criteria for cadmium, copper, lead, and zinc were exceeded at several stormwater-runoff sites (table 23). These concentrations also were high enough to potentially cause health effects in aquatic biota. Copper has been shown to cause sublethal effects on salmon at concentrations as low as 2 µg/L (Baldwin and others, 2003); most concentrations measured in this study were greater than 2 µg/L. Chromium and zinc have been determined to cause reproductive issues in rainbow trout at levels as low as 0.005 and 20 µg/L, respectively (Billard and Roubaud, 1985). All chromium and most zinc concentrations measured in this study (table 22) were higher than these sublethal effects levels. Although fish do not live in these stormwater pipes, the mixing zones where runoff enters their receiving waters are likely inhabited by biota.

Stormwater runoff entering the Willamette River near the St. John's railroad bridge, Portland, Oregon, October 2009.

Table 21.   Polycyclic aromatic hydrocarbons (PAHs) detected in unfiltered stormwater runoff, Columbia River Basin, Washington and Oregon, 2009–10.

[Station names are shown in table 3. Concentrations are in micrograms per liter. Present, presence is verified, but concentration is not quantified; **Abbreviations:** E, estimated; –, not detected]

Analyte	Wenatchee	Richland	Umatilla	The Dalles	Hood River	Portland1	Vancouver1	Vancouver2	Portland2	Willamette1	Willamette2 –Dec	Willamette2 –May	Willamette3	Willamette4	St. Helens	Longview
Acenaphthene	–	–	–	–	–	–	–	Present	Present	–	Present	Present	Present	Present	–	Present
Acenaphthylene	–	–	–	–	–	–	–	–	Present	–	–	–	Present	–	–	Present
Anthracene	–	–	Present	Present	–	–	–	Present	Present	–	Present	Present	–	Present	–	Present
Benzo[a]anthracene	–	–	Present	Present	–	–	–	–	Present	–	E0.25	Present	E0.13	–	–	Present
Benzo[a]pyrene	–	–	Present	Present	–	–	Present	E0.17	E0.24	–	E0.25	Present	E0.19	Present	–	E0.17
Benzo[b]fluoranthene	–	Present	E0.23	Present	–	–	Present	E0.26	E0.24	–	0.39	Present	E0.19	Present	–	E0.23
Benzo[ghi]perylene	Present	–	Present	Present	–	–	Present	E0.20	Present	Present	E0.23	Present	Present	Present	–	Present
Benzo[k]fluoranthene	–	–	Present	Present	–	–	–	–	Present	–	E0.18	Present	–	–	–	Present
2-Chlorophenol	–	–	–	–	–	–	–	–	–	–	Present	Present	–	–	–	–
Chrysene	–	Present	E0.17	Present	–	–	Present	E0.24	E0.24	–	0.36	Present	Present	Present	–	Present
Dibenz[a,h]anthracene	–	–	Present	–	–	–	–	–	Present	–	–	–	–	–	–	–
1,2-Dichlorobenzene	–	–	–	–	–	–	–	–	–	–	–	Present	–	–	–	–
1,4-Dichlorobenzene	–	–	–	Present	–	–	–	–	–	–	–	Present	–	–	–	–
2,4-Dichlorophenol	Present	Present	Present	–	–	–	–	–	–	–	0.80	2.3	–	–	–	–
Diethyl phthalate	Present	–	–	–	–	Present	–	–	–	Present	1.5	Present	–	–	–	–
Dimethyl phthalate	E0.71	–	–	–	–	0.44	–	Present	Present	Present	E0.61	Present	Present	Present	–	–
2,4-Dimethylphenol	Present	Present	Present	–	–	E0.20	Present	–	Present	0.44	2.0	0.38	E0.23	Present	–	–
Di-n-butyl phthalate	–	–	–	–	–	E0.84	E1.8	–	–	–	E1.0	Present	–	–	–	–
Di-n-octyl phthalate	–	–	Present	–	–	–	–	–	Present	–	–	–	–	–	–	–
bis(2-Ethylhexyl) phthalate	Present	–	–	–	–	E3.0	E2.3	2.1	Present	<2.2	E2.9	–	–	–	–	–
Fluoranthene	Present	Present	E0.22	E0.19	–	Present	E0.21	0.35	E0.41	Present	0.78	Present	E0.25	Present	–	0.21
Fluorene	–	–	Present	–	–	–	Present	Present	Present	–	–	–	Present	Present	–	Present
Indeno[1,2,3-cd]pyrene	–	–	Present	Present	–	–	Present	Present	Present	E0.19	E0.19	–	Present	Present	–	Present
Isophorone	Present	Present	Present	–	–	–	E0.15	Present	–	–	–	Present	Present	–	–	–
Naphthalene	–	–	Present	–	–	–	Present	Present	Present	Present	E0.21	Present	Present	Present	–	–
2-Nitrophenol	–	–	–	–	–	–	–	–	–	Present	–	–	–	–	–	–
4-Nitrophenol	<1.3	–	–	Present	–	–	Present	E0.33	Present	E0.34	–	E0.40	–	E0.29	–	–
N-Nitrosodiphenylamine	Present	–	–	–	–	–	Present	Present	Present	E0.31	E0.18	–	–	Present	–	–
Pentachlorophenol	E0.49	–	–	Present	–	E0.57	Present	Present	E0.86	Present	E0.43	E0.50	Present	Present	–	–
Phenanthrene	Present	–	Present	–	–	Present	E0.20	E0.19	E0.19	–	0.65	Present	Present	Present	–	Present
Phenol	Present	Present	Present	–	–	0.37	–	E0.14	Present	0.32	2.1	E0.15	–	Present	–	Present
Pyrene	Present	Present	E0.18	E0.18	–	Present	E0.18	0.36	E0.37	Present	0.63	Present	E0.25	Present	–	E0.19
1,2,4-Trichlorobenzene	–	–	Present	–	–	–	–	–	–	–	–	–	–	–	–	–
2,4,6-Trichlorophenol	–	–	–	–	–	–	–	–	–	–	–	E0.20	–	–	–	–

Table 22.   Trace elements detected in stormwater runoff, Columbia River Basin, Washington and Oregon, 2009–10.

[Station names are shown in table 3. Concentrations are in micrograms per liter. See table A6 for a listing of trace-element compounds analyzed and their reporting limits. Present, presence is verified, but concentration is not quantified. **Abbreviation:** E, estimated; –, not detected]

Analyte	Wenatchee	Richland	Umatilla	The Dalles	Hood River	Portland1	Vancouver1	Vancouver2	Portland2	Willamette1	Willamette2 –Dec	Willamette2 –May	Willamette3	Willamette4	St. Helens	Longview
Unfiltered water																
Arsenic	0.87	2.6	2.3	2.1	Present	0.98	0.64	0.49	0.42	1.1	2.5	2.1	1.8	0.97	0.25	1.7
Cadmium	0.08	–	0.50	0.57	–	0.13	0.21	0.10	0.11	0.17	0.77	0.65	0.08	0.15	Present	E 0.04
Chromium	0.71	0.60	9.0	2.1	–	2.5	6.8	2.9	1.3	2.4	4.7	2.8	4.7	33	Present	0.64
Copper	9.6	6.3	42	7.6	Present	20	15	7.8	12	16	22	15	9.3	12	2.2	3.1
Lead	2.2	0.76	19	1.2	Present	5.8	21	13	3.6	3.2	53	24	5.0	5.7	0.62	1.3
Mercury	–	–	0.09	E 0.01	–	–	0.01	E 0.01	E 0.01	–	0.18	0.07	0.02	0.01	–	–
Nickel	1.1	1.5	22	2.4	0.23	2.2	2.2	1.2	1.2	2.3	4.3	4.0	3.5	3.3	0.63	0.90
Selenium	–	0.73	0.13	0.20	–	Present	–	–	–	–	0.33	0.25	E 0.10	–	Present	–
Silver	–	–	0.13	–	–	0.018	0.019	0.018	Present	Present	0.25	0.078	0.020	0.023	–	–
Zinc	87	17	160	27	7.5	50	93	34	73	150	190	150	28	250	11	28
Filtered water																
Arsenic	0.69	2.3	0.42	2.0	0.13	0.73	0.26	0.37	0.26	0.43	1.1	1.4	0.31	0.56	0.14	0.80
Cadmium	0.08	0.05	0.03	0.6	E 0.02	0.07	0.07	0.07	0.08	0.14	0.25	0.48	Present	0.05	E 0.02	0.02
Chromium	Present	Present	Present	1.1	Present	1.0	1.6	2.1	0.50	0.68	0.62	Present	Present	Present	Present	Present
Copper	9.2	4.5	5.4	7.6	4.6	11	5	7.4	7.8	8.3	5.2	2.6	3.4	1.8	1.9	E 1.0
Lead	0.32	0.19	0.12	0.16	0.11	0.59	0.31	12	0.47	0.18	2.9	2	0.08	0.10	0.08	0.03
Mercury	–	–	–	–	–	–	–	–	–	–	0.01	–	–	–	–	–
Nickel	1.3	0.89	1.4	1.5	0.23	1.1	0.71	0.98	0.86	1.6	1.9	3.2	0.98	1.0	0.70	0.50
Selenium	0.05	0.61	E 0.04	0.20	Present	E 0.04	Present	–	Present	0.05	0.34	0.21	0.06	0.05	Present	Present
Silver	–	–	0.02	–	–	–	–	–	–	E 0.01	0.01	–	–	–	–	–
Zinc	85	12	17	32	7.7	28	39	7.7	60	100	57	47	E 4.5	87	11	18

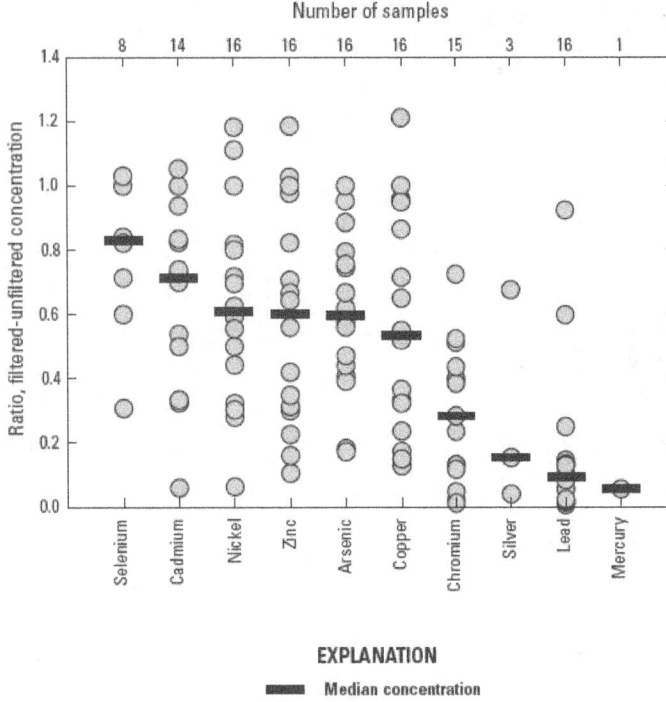

EXPLANATION

▬▬ Median concentration

Figure 4. Ratio of filtered to unfiltered concentrations of trace elements measured in stormwater runoff, Columbia River Basin, Washington and Oregon, 2009–10.

Although mercury was determined as part of the trace-elements analytical suite at NWQL for all stormwater samples, aliquots from stormwater samples collected after November 2009 also were analyzed for mercury and methylmercury in unfiltered water at much lower detection limits at the Wisconsin Mercury Research Laboratory (table 15). The Willamette2 site continued to be a key contributor with the highest, by far, concentrations of all the stormwater samples at 230 ng/L in December and 74 ng/L in May. Although these values seem very high, they are less than the ODEQ-established screening level of 770 ng/L for total mercury in stormwater outfalls in the Portland Harbor area (Oregon Department of Environmental Quality and U.S. Environmental Protection Agency, 2005). The other locations with concentrations greater than or equal to the chronic criteria for freshwater aquatic life (12 ng/L) are Vancouver2, Willamette3, and Willamette4. The levels of methylmercury were all low, except at Willamette2, which was similar in concentration to that measured from The Dalles WWTP.

Stormwater outfall near Interstate Highway 205 bridge (Portland1), Portland, Oregon, October 2009.

Table 23. Concentrations exceeding freshwater-quality criteria for trace elements in stormwater runoff, Columbia River Basin, Washington and Oregon, 2009–10.

[Station names are shown in table 3. Concentrations are in micrograms per liter; **Acute:** Criteria refer to the average concentration for 1 hour. **Chronic:** Criteria refer to the average concentration for 4 days, neither of these criteria should be exceeded more than once every 3 years. All criteria are based on concentrations in filtered water, except for mercury, which is based on concentrations in unfiltered water; a hardness of 50 milligrams per liter was used for criteria calculations; there were no exceedances of criteria in stormwater collected in Richland, Hood River, St Helens, or Longview. References for criteria are: U.S. Environmental Protection Agency (EPA), 2009b; Oregon (OR) Department of Environmental Quality, 2010b; Washington (WA) State Department of Ecology, 2003. **Abbreviations:** CMC, criterion maximum concentration; CCC, criterion continuous concentration. E, estimated]

Analyte	Agency standard	Acute (CMC)	Chronic (CCC)	Concentration range detected in this study	Criterion exceeded	Wenatchee	Umatilla	The Dalles	Portland1	Vancouver1	Vancouver2	Portland2	Willamette1	Willamette2 -Dec	Willamette2 -May	Willamette3	Willamette4
Cadmium	EPA	1.03	0.15	E 0.01–0.6	Chronic									X	X		
Copper	OR, WA	8.86	6.28	E 0.68–11	Acute	X			X				X				
					Chronic	X		X	X				X				
Copper	EPA	6.99	4.95	E 0.68–11	Acute	X		X	X		X	X	X				
					Chronic	X	X	X	X	X	X	X	X				
Lead	OR, WA	30	1.17	0.03–12	Chronic						X			X	X		
Lead	EPA	39	117	0.03–12	Chronic						X			X	X		
Mercury	WA	2.1	0.012	0.002–0.23	Chronic						X			X	X	X	X
Mercury	OR	2.4	0.012	0.002–0.23	Chronic						X	X		X	X	X	X
Zinc	OR, WA	64	58	4.5–100	Acute	X							X				X
					Chronic	X							X				X
Zinc	EPA	65	66	4.5–100	Acute	X							X				X
					Chronic	X							X				X

Oil and Grease

Oil and grease concentrations in this study were consistent and near or less than the reporting limit (table 24). The only location where concentrations were greater than the reporting limit was Willamette2. Analytical difficulties with these analyses caused method blanks often to show concentrations equal to one-half or more of the concentration in the environmental sample.

Synopsis

The overall percentage of compounds detected in the stormwater-runoff samples (58 percent, or 114 of 195, fig. 5) was very similar to the percentage detected in WWTP-effluent samples (53 percent, or 112 of 120, fig. 3). The difference for stormwater is that the compounds detected were not similar across locations. Trace elements were detected at all sites and at levels of concern for the health of aquatic biota. All of the other compound classes were dominated by a few samples with high suspended-sediment concentrations— Umatilla, Willamette3, and Willamette2. The suspended-sediment contribution alone could not account for the large number and elevated concentrations measured at the Willamette2 site. Land-use sources in the drainage area play a key role at this site, which is located within an EPA Superfund project area. Two of the ubiquitous compound classes detected in the stormwater runoff, trace elements and PAHs, are related to automobiles and impervious surfaces, typical findings for stormwater runoff in urban areas.

Table 24.   Oil and grease detected in stormwater runoff, Columbia River Basin, Washington and Oregon, 2009–10.

[Station names are shown in table 3. Concentrations are in milligrams per liter. **Abbreviations:** –, not detected; E, estimated]

Short name	Date	Time	Reporting limit	Concentration in sample	Concentration in method blank
Wenatchee	12-21-09	1340	5.0	–	–
Richland	05-02-09	1200	5.0	E 4.4	E 1.5
Umatilla	10-04-09	0920	5.0	E 4.4	E 2.4
The Dalles	02-23-09	1210	5.0	E 3.5	E 2.1
Hood River	02-23-09	1310	5.0	E 4.1	E 2.1
Portland1	10-14-09	1100	5.0	E 4.8	E 2.3
Vancouver1	12-16-09	1340	5.0	E 2.3	–
Vancouver2	12-16-09	1210	5.0	E 3.3	–
Portland2	10-26-09	1210	5.0	E 3.6	–
Willamette1	06-04-10	0840	4.7	E 4.0	E 3.4
Willamette2–Dec	12-15-09	1330	5.0	5.6	–
Willamette2–May	05-26-10	1310	5.4	6.7	E 3.6
Willamette3	12-15-09	1310	5.0	–	–
Willamette4	05-26-10	1410	5.5	4.1	–
St Helens	03-30-10	1310	4.7	E 2.5	E 1.7
Longview	03-30-10	1410	4.7	E 2.5	E 1.7

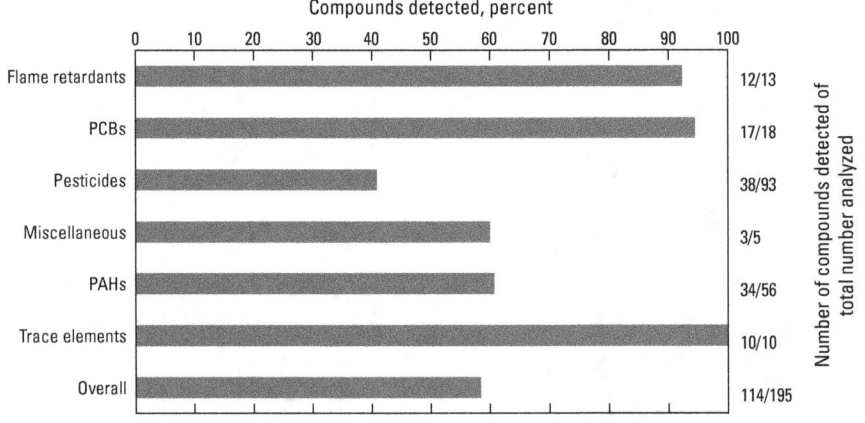

Figure 5.   Percentage of compounds detected in stormwater runoff, Columbia River Basin, Washington and Oregon, 2009–10.

Urban stormwater outfall, Portland, Oregon, October 2009.

Stormwater outfall with spring flowers in downtown Portland, Oregon, May 2009.

Implications for Columbia River Basin

This study was designed as a reconnaissance to inform researchers, scientists, plant operators, policy-makers, and regulators about what compounds are being delivered to the Columbia River through two pathways, WWTP effluent and stormwater runoff. The target analytes included a broad array of compounds, including some of Oregon's SB 737 priority pollutant list as well as emerging contaminants such as endocrine disruptors, personal care products, flame retardants, and other contaminants for which there is little available data for the Columbia River Basin. These data can provide a useful framework for directing future work to identify and reduce contaminant concentrations in the Columbia River Basin.

Four Case Studies

Two pharmaceuticals—one over-the-counter and one prescription—and two AOCs—one musk and one detergent metabolite, both endocrine disruptors—were selected for further study. All four of these compounds were consistently detected in WWTP effluent (fig. 6) and in the Columbia River in previous studies. Diphenhydramine is an over-the-counter antihistamine used to treat the symptoms of hay fever, allergies, and the common cold; to prevent and treat motion sickness; to treat insomnia; and to control abnormal muscle movements in patients with early Parkinson's syndrome (PubMed Health, 2010). Diphenhydramine was detected in samples collected at all nine WWTPs (table 10) with a median concentration of 0.062 µg/L. During a study in the lower Columbia River estuary, a trace-level amount of diphenhydramine was detected in filtered water from the Columbia River at Point Adams near the mouth of the river in April 2005 (Lower Columbia River Estuary Partnership, 2007). At Point Adams, a large amount of water is available for dilution, but diphenhydramine was still detected. Little is known about the effects of diphenhydramine on aquatic biota, but earthworms living in soils treated with biosolids accumulated diphenhydramine (Kinney and others, 2008).

Trimethoprim is an antibiotic prescribed for urinary tract infections and also can be used to treat pneumonia and "traveler's diarrhea" (PubMed Health, 2008). When combined with sulfamethoxazole, trimethoprim is often used to treat ear infections and chronic bronchitis. Besides its use for humans, it also is registered for use in dogs, horses, cattle, and swine. The median concentration of trimethoprim measured in WWTP effluents in this study was 0.089 µg/L, and it was detected in eight of the nine WWTPs sampled (table 10). In August 2004, trimethoprim was detected at low levels in the Columbia River at Warrendale (just downstream of Bonneville Dam), the Willamette River at Portland, and the Columbia

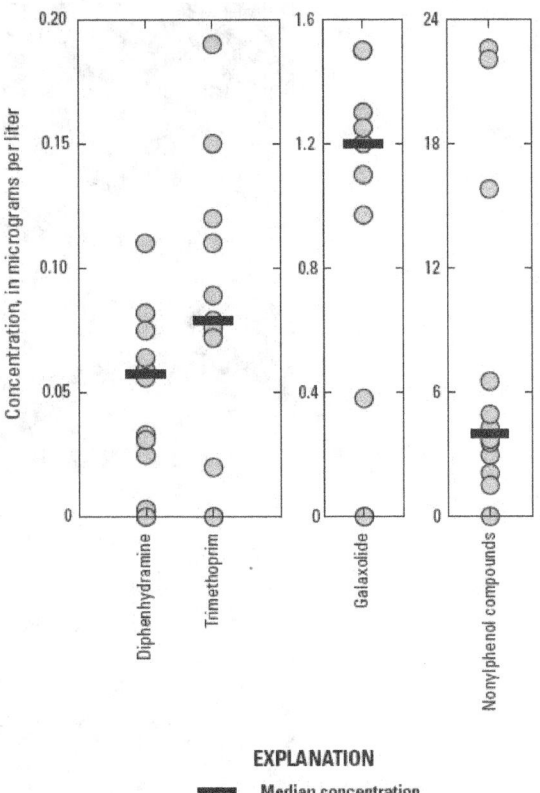

EXPLANATION

▬▬ Median concentration

Figure 6. Concentrations of diphenhydramine, trimethoprim, galaxolide, and nonylphenol compounds in wastewater-treatment-plant-effluent samples, Columbia River Basin, Oregon, 2008–09. (n = 16 for all compounds.)

River at Beaver Army Terminal near Longview. The direct environmental effects of trimethoprim are not known, but the presence of antibiotics in the aquatic environment could lead to microbial resistance (Kümmerer, 2004).

Galaxolide®, the short name for hexahydrohexamethylcyclopentabenzopyran (HHCB), is a synthetic fragrance used in cosmetics, cleaning agents, detergents, air fresheners, and perfumes (International Flavors and Fragrances, Inc., 2007). Galaxolide has been shown to be anti-estrogenic (Schreurs and others, 2005) and bioaccumulate in the food web (Hu and others, 2011). Galaxolide was detected in samples collected at all nine WWTPs (table 9) with a median concentration of 1.2 µg/L. During the lower Columbia River estuary study, trace levels of galaxolide were detected in the Columbia and Willamette Rivers during both high- and low-flow sampling events (Lower Columbia River Estuary Partnership, 2007).

Nonylphenol compounds are a group of nonionic detergent metabolites that are present in cleaners and sealers, and are produced in large quantities in the United States (table A2). They are known endocrine disruptors and are toxic to aquatic life, causing reproductive effects in aquatic organisms (U.S. Environmental Protection Agency, 2006). Nonylphenol compounds are resistant to natural degradation in water and their presence in WWTP effluent is often as a breakdown product from surfactants and detergents. For this discussion, the results for the five nonylphenol variants (*para*-nonylphenol, NP1EO, NP2EO, OP1EO, OP2EO) analyzed are summed to provide a total measure of nonylphenol compounds. These compounds were detected at 8 of the 9 WWTPs (table 9) with a median concentration in Portland of 22 µg/L and a median for the other 7 plants of 3.6 µg/L. From a study of emerging contaminants in the surficial bed sediment of the lower Columbia River and its tributaries in 2005–06, nonylphenol compounds were detected in small tributaries (Fanno Creek), larger tributaries (the Tualatin River and Columbia Slough), and the Columbia River (upstream of the Willamette River and near the Cowlitz River) (Nilsen and others, 2007). Therefore, nonylphenol compounds may have been discharged from WWTPs consistently enough and at high enough concentrations to be measurable in the receiving-water environments.

The freshwater aquatic-life criteria for nonylphenols are 28 µg/L for the acute (1-hour average) criterion and 6.6 µg/L for the chronic criterion (4-day average) (U.S. Environmental Protection Agency, 2006). Therefore, the concentrations measured at the Portland WWTP exceed the chronic criterion, but they do not represent a 4-day average. The European Union has banned nonylphenol and nonylphenol ethoxylates because they have been determined to be a hazard to human and environmental safety (Sierra Club, 2005).

Loadings to the River

For these four example compounds, instantaneous loadings from each WWTP were calculated and then extrapolated to represent a daily load to the Columbia River (table 25). These loads were determined by multiplying the concentration of a given contaminant by the daily discharge for the WWTP (table 12) and a conversion factor. This calculation provides a measure of the instantaneous load of the contaminant entering the river at the point of discharge if it is assumed that the measured concentration is a good representation of the concentration throughout the day for the WWTP effluent. The data necessary to quantify the variability for each contaminant throughout the day at a given WWTP were not collected for this study, but for exploratory purposes, these instantaneous loads can be useful to indicate potentially important sources of contaminants to the Columbia River. The instantaneous loads for the Portland WWTP were consistently higher than for any other wastewater-treatment plant because

the discharge from the Portland WWTP is higher than for any other plant. The discharge from Portland is five times larger than the next largest WWTP in Vancouver. Additionally, the concentrations measured from the Portland WWTP also tended to be higher, particularly for the nonylphenols, indicating that this may be an area to focus future reduction efforts.

The calculations were taken a step further by dividing the instantaneous load by the 7Q10 streamflow of the Columbia River at that point of discharge (table 2) and multiplying by a conversion factor to determine the resulting concentration in the Columbia River that could be attributed to this incoming load (table 25). These calculations illustrate that the Columbia River carries a large amount of water and is able to "absorb" a variety of inputs because of its ability to dilute.

The calculated concentrations were all small, mostly less than 1 ng/L, except near the Portland WWTP. These concentrations, besides Portland, would not be measurable in the Columbia River main stem using standard sampling techniques. Although these calculated concentrations are small in the context of the whole river, the localized effect of these continuous inputs to the mixing zone is understudied and underrepresented. The aquatic biota using these areas may be exposed to higher concentrations than in other areas.

Comparison to SB 737 Plan Initiation Levels

As part of Oregon's SB 737 process to identify persistent pollutants and reduce concentrations entering Oregon's waterways (Oregon Department of Environmental Quality, 2010a), ODEQ was tasked with developing PILs for each of these pollutants. These PILs are used to determine if a city will be required to develop a reduction plan for each persistent pollutant. ODEQ selected these PILs from several existing values, including drinking-water maximum-contaminant levels for those that are established, peer-reviewed national and international government documents, and scientific literature that reflect current scientific information.

The reconnaissance data from this study were compared to the SB 737 list and PILs to provide some preliminary information. Of the 118 persistent pollutants on the SB 737 list, 63 were not analyzed as a part of this study, and 13 were analyzed only in stormwater-runoff samples (table 26). Senate Bill 737 applies only to WWTP effluent and not to stormwater runoff. Of the remaining 42 pollutants analyzed in wastewater, 27 were detected. Only 4 of the 27 were measured at least once at a level greater than the assigned PIL (table 27). One of these, fluoranthene, was detected only in the sample from the Wenatchee WWTP (E 0.11 µg/L), which is in Washington and, therefore, not subject to the requirements of Oregon SB 737. Another PAH, anthracene, was detected in a sample collected at the St. Helens WWTP at greater than the PIL of 0.01 µg/L. It was measured at a level less than the LT-MDL (reported as "Present"), but the quantified result slightly exceeded the PIL of 0.01 µg/L.

Table 25. Instantaneous loadings and calculated concentrations for diphenhydramine, trimethoprim, galaxolide, and nonylphenol compounds in the Columbia River, Columbia River Basin, Washington and Oregon, 2008–09.

[Station names are shown in table 2. A 7Q10 flow was not available for Umatilla, so 78,000 ft^3/s was used for these calculations **Abbreviatons:** Mgal/d, million gallons per day; 7Q10, the lowest streamflow for 7 consecutive days that occurs on average once every 10 years; this value is the Columbia River flow near the discharge point that is used for calculating mixing zones; ft^3/s, cubic foot per second; µg/L, micrograms per liter; g/d, grams per day]

Analyte	Wenatchee (2008)	Wenatchee (2009)	Richland	Umatilla	The Dalles	Hood River	Portland (a.m.)	Portland (noon)	Portland (p.m.)	Vancouver	St. Helens	Longview
Flow values												
Daily discharge for sampling date (Mgal/d)	3.1	2.9	5.4	0.54	1.7	0.89	49	49	49	10	6.9	6.9
Columbia River 7Q10 flow (ft^3/s)	51,557	51,557	52,700	78,000	80,637	74,000	79,436	79,436	79,436	79,436	88,900	97,400
Diphenhydramine												
Concentration in sample (µg/L)	0.090	0.059	0.025	Present	0.11	0.082	0.075	0.064	0.056	0.10	0.033	0.031
Instantaneous load (g/d)	1.1	0.6	0.5	—	0.7	0.3	14	10	10	3.8	0.9	1
Calculated concentration in Columbia River (µg/L)	0.00001	0.00001	0.000004	—	0.000004	0.000002	0.0001	0.0001	0.0001	0.00002	0.000004	0.000004
Trimethoprim												
Concentration in sample (µg/L)	0.15	0.19	0.1	—	0.11	0.12	0.089	0.079	0.076	0.073	0.02	0.072
Instantaneous load (g/d)	1.8	2.1	2.0	—	0.7	0.4	17	15	14	2.8	0.5	1.9
Calculated concentration in Columbia River (µg/L)	0.00001	0.00002	0.00002	—	0.000004	0.000002	0.00008	0.00008	0.00007	0.00001	0.000002	0.000008
Galaxolide												
Concentration in sample (µg/L)	1.5	2.5	1.5	1.3	1.2	1.1	1	1.2	0.97	1.3	0.38	1.2
Instantaneous load (g/d)	18	27	31	2.7	7.7	3.7	190	220	180	47	10	31
Calculated concentration in Columbia River (µg/L)	0.0001	0.0002	0.0002	0.00001	0.00004	0.00002	0.0010	0.0011	0.0009	0.0002	0.00005	0.0001
Nonylphenol compounds												
Concentration in sample (µg/L)	3.0	2.1	4.3	—	6.5	3.6	23	22	16	1.5	4.9	3.7
Instantaneous load (g/d)	35	23	8.8	—	40	12	4,200	4,100	2,900	58	130	97
Calculated concentration in Columbia River (µg/L)	0.0003	0.0002	0.00005	—	0.0002	0.00007	0.02	0.02	0.01	0.0003	0.0006	0.0004

Table 26. Oregon Senate Bill 737 priority persistent pollutants not detected in wastewater-treatment-plant (WWTP) effluent, Columbia River Basin, Washington and Oregon, 2008–09.

[**CAS No.:** Chemical Abstract Service Registry Number® is a Registered Trademark of the American Chemical Society. CAS recommends the verification of the CASRNs through CAS Client Services[SM]. **Abbreviation:** μg/L, micrograms per liter]

CAS No.	Chemical name	Plan initiation level (μg/L)	CAS No.	Chemical name	Plan initiation level (μg/L)
Analyzed in WWTP effluent but not detected			Not analyzed in this study—Continued		
50-32-8	Benzo[a]pyrene	0.2	1024-57-3	Heptachlor epoxide	0.2
91465-08-6	lambda-Cyhalothrin	0.01	32241-08-0	Heptachloronaphthalene	0.4
72-54-8	p,p'-DDD	0.1	25637-99-4	Hexabromocyclodecane (HBCD)	7
72-55-9	p,p'-DDE	0.1	319-84-6	alpha-Hexachlorocyclohexane	0.006
50-29-3	p,p'-DDT	0.001	319-85-7	beta-Hexachlorocyclohexane	0.04
333-41-5	Diazinon	0.2	58-89-9	gamma-Hexachlorocyclohexane (Lindane)	0.2
1031-07-8	Endosulfan sulfate	0.1	1335-87-1	Hexachloronaphthalene	1.4
88671-89-0	Myclobutanil	200	70-30-4	Hexachlorophene	2
27304-13-8	Oxychlordane (single isomer)	0.4	465-73-6	Isodrin	0.6
42874-03-3	Oxyfluorfen	1.3	330-55-2	Linuron	0.09
35693-99-3	PCB-52	0.5	832-69-9	1-Methylphenanthrene	0.7
31508-00-6	PCB-118	0.5	2381-21-7	1-Methylpyrene	20
35065-28-2	PCB-138	0.5	2385-85-5	Mirex	0.001
40487-42-1	Pendimethalin	6	15323-35-0	Musk indane	10
82-68-8	Pentachloronitrobenzene	20	81-14-1	Musk ketone	30
Analyzed in stormwater runoff only			145-39-1	Musk tibetene	4
			81-15-2	Musk xylene	100
7440-38-2	Arsenic compounds (in filtered effluent)	10	1163-19-5	PBDE-209	0.1
56-55-3	Benz[a]anthracene	0.02	7012-37-5	PCB-28	0.5
205-99-2	Benzo[b]fluoranthene	0.5	32598-13-3	PCB-77	0.5
191-24-2	Benzo[ghi]perylene	2	70362-50-4	PCB-81	0.5
207-08-9	Benzo[k]fluoranthene	0.002	32598-14-4	PCB-105	0.5
7440-43-9	Cadmium compounds (in filtered effluent)	5	74472-37-0	PCB-114	0.5
218-01-9	Chrysene	2	65510-44-3	PCB-123	0.5
53-70-3	Dibenz[ah]anthracene	0.0004	57465-28-8	PCB-126	0.5
193-39-5	Indeno[1,2,3-cd]pyrene	0.5	35065-27-1	PCB-153	0.5
7439-92-1	Lead compounds (in filtered effluent)	15	38380-08-4	PCB-156	0.5
29082-74-4	Octachlorostyrene	0.2	69782-90-7	PCB-157	0.5
7782-49-2	Selenium compounds (in unfiltered effluent)	50	52663-72-6	PCB-167	0.5
88-06-2	2,4,6-Trichlorophenol	2	32774-16-6	PCB-169	0.5
Not analyzed in this study			39635-31-9	PCB-189	0.5
			608-93-5	Pentachlorobenzene	6
98-07-7	Benzotrichloride (trichloromethylbenzene)	0.03	1321-64-8	Pentachloronaphthalene	4
82657-04-3	Bifenthrin	0.02	375-85-9	Perfluoroheptanoic acid (PFHpA)	300
56-35-9	bis-(Tributyltin) oxide (TBTO)	0.008	375-95-1	Perfluorononanoic acid (PFNA)	1
143-50-0	Chlordecone (Kepone)	0.5	754-91-6	Perfluorooctane sulfonamide (PFOSA)	0.2
541-02-6	Cyclopentasiloxane, decamethyl- (D5)	16	1763-23-1	Perfluorooctane sulfonic acid (PFOS)	300
556-67-2	Cyclotetrasiloxane, octamethyl- (D4)	7	335-67-1	Perfluorooctanoic acid (PFOA)	24
434-90-2	Decafluorobiphenyl	18	2062-78-4	Pimozide	3
52918-63-5	Deltamethrin (decamethrin)	0.0004	67747-09-5	Prochloraz	2
115-32-2	Dicofol	6	80214-83-1	Roxithromycin	710
56-53-1	Diethylstilbestrol	87	92-94-4	p-Terphenyl	11
88-85-7	Dinoseb	7	79-94-7	Tetrabromobisphenol A (TBBPA)	980
1746-01-6	Dioxins/furans (as 2,3,7,8-TCDD TEQ)	0.00003	1335-88-2	Tetrachloronaphthalene	14
72-20-8	Endrin	2	1321-65-9	Trichloronaphthalene	43
66230-04-4	Esfenvalerate	0.02	95-95-4	2,4,5-Trichlorophenol	18
13356-08-6	Fenbutatin-oxide	0.5	732-26-3	2,4,6-Tris-(1,1-dimethylethyl)phenol	6
76-44-8	Heptachlor	0.4			

Table 27.   Oregon Senate Bill 737 priority persistent pollutants detected in wastewater-treatment-plant (WWTP) effluent, Columbia River Basin, Washington and Oregon, 2008–09.

[**CAS No.:** Chemical Abstract Service Registry Number® is a Registered Trademark of the American Chemical Society. CAS recommends the verification of the CASRNs through CAS Client Services[SM]. **Plan initiation level (PIL; SB 737):** From Oregon Department of Environmental Quality (2010a). **Abbreviatons:** μg/L, micrograms per liter; E, estimated]

CAS No.	Chemical name	Plan initiation level (μg/L)	Number of WWTPs with detections	Number of WWTPs with detections greater than PIL	Range of detections (μg/L)	WWTPs with detections
\multicolumn{7}{c}{Detected in WWTP effluent at concentrations greater than the PIL}
57-88-5	Cholesterol	0.06	9	9	Present–E 6.3	see table 9
360-68-9	Coprostanol	0.04	9	9	Present–E 5.8	see table 9
120-12-7	Anthracene	0.01	1	1	Present	St. Helens, see table 9
206-44-0	Fluoranthene	0.04	1	1	E 0.11	Wenatchee, see table 9
\multicolumn{7}{c}{Detected in WWTP effluent at concentrations less than the PIL}
5103-71-9	*cis*-Chlordane	2	9	0	0.00002–0.00019	see table 13
5103-74-2	*trans*-Chlordane	2	9	0	0.00001–0.00019	see table 13
2921-88-2	Chlorpyrifos	0.04	4	0	0.0000–0.00043	see table 13
60-57-1	Dieldrin	0.002	5	0	0.00001–0.00017	see table 13
120068-37-3	Fipronil	15	7	0	Present–E 0.13	see tables 13 and 14
1222-05-5	Galaxolide (HHCB)	29	9	0	0.38–2.5	see table 9
118-74-1	Hexachlorobenzene (HCB)	1	1	0	Present	Umatilla, see table 13
22967-92-6	Methylmercury	0.004	7	0	0.00006–0.0004	see table 15
5103-73-1	*cis*-Nonachlor	2	2	0	Present –0.00001	see table 13
39765-80-5	*trans*-Nonachlor	2	8	0	Present–0.0001	see table 13
5436-43-1	PBDE-47	0.7	9	0	Present	see table 13
60348-60-9	PBDE-99	0.7	9	0	Present–0.0048	see table 13
189084-64-8	PBDE-100	0.7	9	0	Present–0.0016	see table 13
68631-49-2	PBDE-153	1	9	0	0.00006–0.0004	see table 13
37680-73-2	PCB-101	0.5	1	0	Present	Longview, see table 13
35065-29-3	PCB-180	0.5	3	0	0.00001–0.00002	see table 13
1825-21-4	Pentachloroanisole	35	5	0	Present –0.00085	see table 13
85-01-8	Phenanthrene	0.4	1	0	Present	St. Helens, see table 9
129-00-0	Pyrene	0.03	3	0	Present	see table 9
83-46-5	*beta*-Sitosterol	25	9	0	E 0.79–E 3.2	see table 9
83-45-4	*beta*-Stigmastanol	75	5	0	Present–E 1.1	see table 9
3380-34-5	Triclosan	70	8	0	Present–0.76	see tables 9 and 13
1582-09-8	Trifluralin	1.1	3	0	Present–0.00002	see table 13

The other two compounds from the SB 737 list that were detected in this study at concentrations that were greater than the PILs are both sterols that are ubiquitous in wastewater. Cholesterol is a structural component of mammalian cell walls and is transported in the blood plasma of all animals. Its effects in the aquatic environment are not understood. Coprostanol is a metabolite of cholesterol excreted in fecal matter from mammals and is, therefore, detected in urban effluents. It can be accumulated by organisms living in municipal effluent outfalls and has been determined to have estrogenic effects in freshwater mussels (Gagné and others, 2001). According to ODEQ, "Research suggests that coprostanol has an affinity to estradiol-binding sites and … large amounts may pose a threat to aquatic invertebrates" (Oregon Department of Environmental Quality, 2010a).

A large number of the AOCs and pharmaceuticals analyzed in WWTP-effluent samples in this study were detected, yet only 27 of the 42 compounds on the SB 737 persistent pollutants list that were analyzed in this study were detected in these WWTP-effluent samples. The reason for this dichotomy is that many of the contaminants on the persistent pollutant list are PAHs, metals, and currently used pesticides—all compounds that are likely to be detected in stormwater but not necessarily wastewater. To illustrate this point, 45 of the 50 compounds on the persistent pollutant list that were analyzed in stormwater-runoff samples in this study were detected; however, SB 737 does not apply to stormwater. One category that is absent from the persistent pollutant list is pharmaceuticals, the contaminant class that many people associate with WWTP effluent. Sufficient documentation about the persistence or bioaccumulative ability of many pharmaceuticals is not available to allow consideration for the persistent pollutant list.

Future Directions

This study was intended to serve as a stepping-stone for future work. Therefore, it is important to consider how the data from this reconnaissance can help inform decisions about sampling design and identify information gaps. This type of information can be combined to more effectively address potential reduction efforts depending on the targeted contaminant class.

Sampling Design

The samples collected in this study were point or grab samples designed for an initial characterization of the pathways sampled. Through these efforts, it was determined that WWTP effluent contains a wide variety of contaminants from many different compound classes. No clear pattern of detections emerged among the WWTPs based on location, population, treatment type, or size of the plant. This type of ancillary information could not be used to anticipate the type or frequency of detections. Given the various factors influencing the composition of the effluent, it would be difficult to design a study to explain the expected results for WWTP effluent. This pathway can be considered simply as an integrator of human activity and used to evaluate the effects this activity has on the ecosystem.

At the Portland WWTP, three samples were collected throughout 1 day to examine temporal variability. Given the inherent variability in these types of samples, no obvious differences throughout the day were noted for most compounds. The exception was some of the halogenated compounds (PBDEs, herbicides and insecticides, and triclosan) that exhibited patterns of lower concentrations in the morning with some noon and afternoon concentrations roughly two to four times higher (table 13). This finding illustrates that a single grab sample may not be adequate to characterize the effluent. Time-composited (24-hour) samples may provide a better characterization of the contributions throughout the day.

Another sampling technique that provides a time-integrated sample is the use of passive samplers (Alvarez, 2010). Passive samplers left in the water for about 30 days integrate the sample by accumulating contaminants, similar to fish or other aquatic biota in the area. In this way, these samples provide a biologically relevant measure of the complex mixtures present. Additionally, because these samplers integrate over time, the contaminants are essentially concentrated into the sampling media, which results in increased sensitivity and lower detection limits than a traditional water sample. This would be important for collecting samples not only from the effluent stream but also from the receiving waters, where dilution makes it difficult to detect these low concentrations using standard methods (table 25). An ideal design for future characterization of not only what is being delivered by the WWTP effluent or stormwater runoff, but also the fate and transport of these contaminants, would combine passive samplers in the waste stream and the receiving water, both in the mixing zone and at some distance downstream and outside the mixing zone.

Seasonality is a key factor in characterizing stormwater runoff. The length of dry time preceding a runoff event, the magnitude and length of the event, and the time of year (related to pesticide usage, for example) all may have an effect (Sansalone and others, 1998; Flint, 2004). Much less is known about seasonality effects on WWTP effluent. Much will depend on the type of inputs the WWTP receives. For instance, a WWTP receiving wastewater from a cannery or fruit-processing facility will see changes in their influent as the contributing facilities change their operations. How much of an effect this has on the makeup of the effluent could be a study objective. Likewise, seasonal changes in prescription and over-the-counter medicine use also may be noticeable in WWTP effluents.

The chemical characteristics of the targeted compounds often determine how to focus the sampling media. For hydrophobic compounds, sampling should be targeted at retrieving as much solid material as possible, either through sampling unfiltered water or the solids themselves. The dissolved phase should not be ignored, however, because most compounds are not confined to just one media. The samples measured for estrogenicity in this study indicated that the hormonally active compounds were likely dissolved in the effluent water rather than associated with the solids in the effluent (table 11).

Information Gaps

The affinity that some contaminants have for the solid phase rather than the aqueous phase raises questions about how many contaminants are sequestered in biosolids during the treatment process and what is their ultimate fate. For most WWTPs, biosolids are transported to a nearby area and spread over the land, many times applied as a fertilizer or nutrient treatment for the land. In 2008, USGS researchers determined that earthworms collected from fields where biosolids and manure had been applied not only picked up drugs and perfumes from the soil but also had bioconcentrated these compounds in their tissues (Kinney and others, 2008). Once these contaminants are spread on the land with the biosolids, little is known about the further transport, degradation, or ultimate effect of these contaminants on the ecosystem.

This study focused on two known pathways for contaminant transport, but further study could expand this focus to incorporate other potential pathways and sources. These could include but are not limited to other dischargers with an NPDES permit; industries located along a contributing stream; hospitals, nursing homes, hospice and in-home facilities; or manufacturers of emerging contaminants. Further characterization of the land-use types and contributions to stormwater-runoff areas also would help in designing reduction efforts based on these results. The spatial distribution of this study could be expanded to include a further extent of the Columbia River Basin, including pathways to tributaries.

The aquatic communities living in these receiving waters are exposed to complex mixtures of these compounds with unknown toxicity. Further research into sublethal effects, toxicity, bioaccumulation, and mixture effects is needed to further the science of emerging contaminants.

Summary and Conclusions

Toxic contamination is a significant concern in the Columbia River Basin. Many efforts and dollars are focused on restoring critical habitat for endangered salmonids and other wildlife that depend on the ecosystem; however, although physical habitat is a prime consideration in restoration decisions, water-quality concerns, specifically contamination issues, also can influence these decisions. Toxics-reduction efforts are underway to protect the health of people, aquatic life, and the ecosystem.

To successfully reduce toxics and restore critical habitat, an understanding of the sources of contaminants is necessary. This study was designed to take a first look at two easily defined pathways that deliver contaminants to the Columbia River, wastewater-treatment-plant (WWTP) effluent and stormwater runoff. The resulting data can be used to assess the types, number, and magnitude of compounds present and to lay the foundation for additional studies and potential toxics-reduction activities.

Nine cities were selected from throughout the Columbia River Basin to provide diversity in physical setting, climate characteristics, and population density. In downstream order, the cities sampled were Wenatchee, Richland, Umatilla, The Dalles, Hood River, Portland, Vancouver, St. Helens, and Longview throughout Washington and Oregon. These cities also were selected because their WWTP effluent and at least some part of their stormwater (except Umatilla) is delivered directly to the Columbia River. Most samples at the WWTPs were collected in December 2008 for anthropogenic organic compounds (AOCs), pharmaceuticals, estrogenicity, and halogenated compounds. In December 2009, each of these WWTPs was revisited to collect samples for the analysis of currently used pesticides, mercury, and methylmercury. Stormwater samples were collected throughout spring and winter storms of 2009 and 2010 from these cities as well as additional sites along the lower Willamette River near downtown Portland. These samples were analyzed for currently used pesticides, halogenated compounds, mercury, polycyclic aromatic hydrocarbons (PAHs), trace elements, and oil and grease.

WWTP effluents—Flame retardants (polybrominated diphenyl ethers [PBDEs] and others) and steroids were consistently detected in WWTP-effluent samples, whereas few pesticides or PAHs were detected, except at Longview. Concentrations of PBDEs were detected at all sites, and the highest concentrations detected were for congeners PBDE-47, PBDE-99, and PBDE-100. No PCBs were detected at most WWTPs, except Wenatchee. Longview also was notable because it had the greatest number of detections and the concentrations were usually among the highest, particularly for the personal-care-product compounds. Fourteen human-health pharmaceuticals were analyzed for and all but albuterol and warfarin were detected in at least one city. Two pharmaceuticals were detected at all of the W sampled, carbamazepine and diphenhydramine. The yeast estrogen screen, an assay that measures the potential biological effects of the mixture of chemicals present in a sample, was used to screen each sample for total estrogenicity. The estrogenicity levels measured in this study were well above levels that have been shown to cause effects in aquatic biota.

Few currently used pesticides were detected in WWTP-effluent samples. The primary compounds detected were fipronil and its degradates, which were in samples collected from all WWTPs except Wenatchee. Fipronil is an insecticide used to control common household pests like ants, beetles, cockroaches, and other insects, and can be in topical pet-care products used to control fleas. The highest total mercury concentrations were measured at The Dalles and Vancouver. Both of these concentrations were greater than 12 ng/L, the chronic criterion for freshwater aquatic life.

Stormwater-runoff—Diverse sources of stormwater runoff and the larger amount of suspended sediment present in these samples relative to that in WWTP-effluent samples resulted in very different results for the stormwater-runoff samples. Additionally, localized sources contributed to the detection patterns observed in these samples. Of the 49 halogenated compounds detected in stormwater-runoff samples, 45 were detected in the Willamette2-Dec sample, which is within the Portland Harbor Superfund area. The PBDE concentrations at Willamette2 were roughly double those in the Umatilla sample and the PCB concentrations at Willamette2 were 20–300 times greater than PCB concentrations in any other stormwater-runoff samples. Herbicide and insecticide detections in solids filtered from stormwater runoff also follow a pattern of high contaminant concentrations in samples with high suspended-sediment concentrations—particularly from Umatilla, Vancouver, and Willamette2. Detections for several pesticides and PCBs from the Willamette2 site in December and May exceeded chronic freshwater-quality criteria. Although many of these concentrations are low (less than 1 microgram per liter), mixtures of some of these pesticides have been determined to have synergistic and additive effects on salmon health when they occur together.

The 10 trace elements measured in filtered and unfiltered stormwater runoff in this study were detected consistently through all samples. Arsenic, cadmium, copper, nickel, selenium, and zinc tended to transport more readily in the dissolved phase, whereas chromium, lead, mercury, and silver were more often detected in the solid phase. Chronic and sometimes acute freshwater-quality criteria for cadmium, copper, lead, and zinc were all exceeded in several stormwater-runoff samples. These concentrations, particularly for copper, chromium, and zinc, also were potentially high enough to cause health effects in aquatic biota. The Willamette stormwater-runoff sites in the Portland Harbor area, as well as Vancouver2, had concentrations of total mercury greater than or equal to the chronic criterion for freshwater aquatic life (12 ng/L).

Implications for the Columbia River Basin—Instantaneous loadings were calculated for four compounds detected in WWTP-effluent samples—diphenhydramine, trimethoprim, Galaxolide, and nonylphenol compounds—to estimate the potential contribution to the Columbia River from the studied WWTPs. The instantaneous loads for the Portland WWTP were consistently much greater than for any other plant because the effluent discharge from the Portland WWTP is much greater than for any other plant, discharging five times more than the next largest WWTP, in The Dalles. The resulting concentrations in the Columbia River from these contributions were calculated. Most of these concentrations were small and would not be detectable using standard sampling techniques. These calculations illustrate that the Columbia River is able to "absorb" a variety of inputs because of its ability to dilute. Nonetheless, although the calculated concentrations are small in the context of the entire river, the local effect of these continuous inputs in the mixing zone is understudied. The aquatic biota inhabiting these areas may be exposed to higher concentrations than in other areas.

Comparison to Oregon Senate Bill (SB) 737—As part of the SB 737 process to identify persistent pollutants and reduce concentrations entering Oregon's waterways, the Oregon Department of Environmental Quality (ODEQ) was tasked with developing a list of persistent pollutants that have a documented effect on human health, wildlife, and aquatic life. The 52 largest WWTPs in Oregon sent samples to the ODEQ laboratory to be analyzed for these pollutants, and the results were compared to plan initiation levels (PILs) developed to decide when action is required to reduce the presence of these pollutants in the effluents. Only four compounds analyzed for in this study—fluoranthene, anthracene, cholesterol, and coprostanol—exceeded the corresponding PILs. Many of the AOCs and pharmaceuticals analyzed in WWTP-effluent samples collected during this study were detected; however, these samples contained only 27 of the 42 compounds on the SB 737 persistent pollutants list that were analyzed for in this study. The reason for this dichotomy is that many of the contaminants on the persistent pollutant list are PAHs, metals, and currently used pesticides—all compounds that are likely to be detected in stormwater but not necessarily wastewater.

Future directions—This study was intended to serve as a precursor for future work. The study results show that WWTP effluent contains a wide variety of contaminants from many compound classes. Given the variety of factors influencing the composition of the effluent, it would be difficult to design a study to explain the expected results for WWTP effluent. It is preferable to consider this pathway simply as an integrator of human activity and focus on minimizing the effects it has on the ecosystem. Seasonality was not addressed in this study design. The large number of hydrophobic compounds that were detected in these effluents indicates that the biosolids from these WWTPs may be potentially significant sources of these contaminants to the ecosystem.

Stormwater runoff acts as an integrator of human activities and can be a source of various compounds to aquatic ecosystems. The inputs from stormwater runoff are more sporadic than the continual input of WWTP effluents, but their potentially large contributions during short periods can still have an effect on biota that inhabit mixing zones in the receiving waters. Toxics-reduction efforts will be more effective when contaminant occurrence and distribution data are coupled with land-use information from the stormwater catchments that drain to the Columbia River.

Data from this study and others like it can provide a useful framework for directing future work on identifying and reducing contaminant concentrations in the Columbia River Basin. Wastewater-treatment plant effluent and stormwater runoff are two pathways for contaminants to reach the receiving waters, but there are other understudied pathways. The results from this study provide a starting point for future work to continue understanding the presence of contaminants in the environment, develop research to characterize the effects of these contaminants on aquatic life, and prioritize future toxic-reduction efforts.

Acknowledgments

Special thanks to (1) each city mentioned for allowing access to the WWTP effluent and for assistance in locating stormwater-runoff locations; (2) Emily Rauch, Brian Cimyotti, and Rachael Pecore for volunteering and helping to collect stormwater samples in remote locations; (3) Lauren Goldberg and Brett VandenHeuvel (Columbia Riverkeeper) and Mark Riskedahl (Northwest Environmental Defense Center) for their support of this project and help gathering valuable information; and (4) Tiffany Rae Jacklin, Daniel Wise, Esther Duggan, Whitney Temple, Henry Johnson, and David Piatt (all USGS) for their help with sampling and processing.

References Cited

Alvarez, D.A., 2010, Guidelines for the use of the semipermeable membrane device (SPMD) and the polar organic chemical integrative sampler (POCIS) in environmental monitoring studies: U.S. Geological Survey Techniques and Methods 1–D4, 28 p. (Also available at http://pubs.er.usgs.gov/publication/tm1D4.)

Alvarez, D.A., Cranor, W.L., Perkins, S.D., Clark, R.C., and Smith, S.B., 2008, Chemical and toxicological assessment of organic contaminants in surface water using passive samplers: Journal of Environmental Quality, v. 37, p. 1,024–1,033.

Arctic Monitoring and Assessment Programme and Arctic Council Action Plan to Eliminate Pollution of the Arctic, 2005, Brominated flame retardants in the Arctic: Arctic Monitorial and Assessment Programme Fact Sheet, 4 p.

Baldwin, D.H., Sandahl, J.F., Labenia, J.S., and Scholz, N.L., 2003, Sublethal effects of copper on coho salmon—Impacts on nonoverlapping receptor pathways in the peripheral olfactory nervous system: Environmental Toxicology and Chemistry, v. 22, no. 10, p. 2,266–2,274.

Berninger, J.P., Du, B., Connors, K.A., Eytcheson, S.A., Kolkmeier, M.A., Prosser, K.N., Valenti, T.W., Jr., Chambliss, C.K., and Brooks, B.W., 2011, Effects of the antihistamine diphenhydramine on selected aquatic organisms: Environmental Toxicology and Chemistry, v. 30, p. 2,065–2,072.

Billard, R., and Roubaud, P., 1985, The effect of metals and cyanide on fertilization in rainbow trout (*Salmo gairdneri*): Water Research, v.19, no. 2, p. 209–214.

Bonn, B.A., 2008, Using the U.S. Geological Survey National Water Quality Laboratory LT-MDL to evaluate and analyze data: U.S. Geological Survey Open-File Report 2008–1227, 73 p. (Also available at http://pubs.usgs.gov/of/2008/1227/.)

Boyd, G.R., Palmeri, J.M., Zhang, S., and Grimm, D.A., 2004, Pharmaceutical and personal care products and endocrine disrupting chemicals in stormwater canals and Bayou St. John in New Orleans, Louisiana, U.S.: Science of the Total Environment, v. 333, p. 137–148.

Boyd, R.A., and Furlong, E.T., 2002, Human-health pharmaceutical compounds in Lake Mead, Nevada and Arizona, and Las Vegas Wash, Nevada, October 2000–August 2001: U.S. Geological Survey Open-File Report 02–385, 18 p. (Also available at http://pubs.usgs.gov/of/2002/ofr02385/.)

Chevrier, J., Harley, K.G., Bradman, A., Gharbi, M., Sjodin, A., and Eskenazi, B., 2010, Polybrominated diphenyl ether (PBDE) flame retardants and thyroid hormone during pregnancy: Environmental Health Perspectives, doi: 10.1289/ehp.1001905, accessed March 2, 2012, at http://ehp03.niehs.nih.gov/article/info:doi/10.1289/ehp.1001905.

Colman, J.R., Baldwin, D., Johnson, L.L., and Scholz, N.L., 2009, Effects of the synthetic estrogen, 17α-ethinylestradiol, on aggression and courtship behavior in male zebrafish (*Danio rerio*): Aquatic Toxicology, v. 91, p. 345–354.

Daughton, C.G., and Ternes, T.A., 1999, Pharmaceuticals and personal care products in the environment—Agents of subtle change: Environmental Health Perspectives, v. 107, supplement 6, p. 907–938.

Davis, A., Shokouhian, M., and Ni, S., 2001, Loading estimates of lead, copper, cadmium and zinc in urban runoff from specific sources: Chemosphere, v. 44, no. 5, p. 997–1,009.

DeWild, J.F., Olson, M.L., and Olund, S.D., 2002, Determination of methyl mercury by aqueous phase ethylation, followed by gas chromatographic separation with cold vapor atomic fluorescence detection: U.S. Geological Survey Open-File Report 01–445, 14 p. (Also available at http://pubs.usgs.gov/of/2001/ofr-01-445/.)

Drevnick, P.E., and Sandheinrich, M.B., 2003, Effects of dietary methylmercury on reproductive endocrinology of fathead minnows: Environmental Science and Technology, v. 37, no. 19, p. 4,390–4,396.

Ferrari, B., Mons, R., Vollat, B., Fraysse, B., Paxeus, N., Giudice, R.L., Pollio, A., and Garric, J., 2004, Environmental risk assessment of six human pharmaceuticals—Are the current environmental risk assessment procedures sufficient for the protection of the aquatic environment?: Environmental Toxicology and Chemistry, v. 23, p. 1,344–1,354.

Fishman, M.J., ed., 1993, Methods of analysis by the U.S. Geological Survey National Water Quality Laboratory—Determination of inorganic and organic constituents in water and fluvial sediments: U.S. Geological Survey Open-File Report 93–125, 217 p. (Also available at http://pubs.er.usgs.gov/publication/ofr93125.)

Fishman, M.J., and Friedman, L.C., 1989, Methods for determination of inorganic substances in water and fluvial sediments: U.S. Geological Survey Techniques of Water-Resources Investigations, book 5, chap. A1, 545 p. (Also available at http://pubs.usgs.gov/twri/twri5-a1/pdf/twri_5-A1_b.pdf.)

Flint, K.R., 2004, Water quality characterization of highway stormwater runoff from an ultra urban area: College Park, University of Maryland, Master's thesis, 209 p., accessed March 2, 2012, at http://drum.lib.umd.edu/bitstream/1903/210/1/umi-umd-1284.pdf.

Furlong, E.T., Werner, S.L., Anderson, B.D., and Cahill, J.D., 2008, Methods of analysis by the U.S. Geological Survey National Water Quality Laboratory—Determination of human-health pharmaceuticals in filtered water by chemically modified styrene-divinylbenzene resin-based solid-phase extraction and high-performance liquid chromatography/mass spectrometry: U.S. Geological Survey Techniques and Methods, book 5, chap. B5, 56 p. (Also available at http://pubs.usgs.gov/tm/tm5b5/.)

Gagné, F., Blaise, C., Lachance, B., Sunahara, G.I., and Sabik, H., 2001, Evidence of coprostanol estrogenicity to the freshwater mussel *Elliptio complanata*: Environmental Pollution, v. 115, p. 97–106.

Garbarino, J.R., and Damrau, D.L., 2001, Methods of analysis by the U.S. Geological Survey National Water Quality Laboratory—Determination of organic plus inorganic mercury in filtered and unfiltered natural water with cold vapor-atomic fluorescence spectrometry: U.S. Geological Survey Water-Resources Investigations Report 01–4132, 16 p. (Also available at http://nwql.usgs.gov/Public/pubs/WRIR01-4132.pdf.)

Garbarino, J.R., Kanagy, L.K., and Cree, M.E., 2006, Determination of elements in natural-water, biota, sediment and soil samples using collision/reaction cell inductively coupled plasma-mass spectrometry: U.S. Geological Survey Techniques and Methods, book 5, sec. B, chap. 1, 88 p. (Also available at http://pubs.usgs.gov/tm/2006/tm5b1/.)

Garbarino, J.R., and Struveski, T.M., 1998, Methods of analysis by the U.S. Geological Survey National Water Quality Laboratory—Determination of elements in whole-water digests using inductively coupled plasma-optical emission spectrometry and inductively coupled plasma-mass spectrometry: U.S. Geological Survey Open-File Report 98–165, 101 p. (Also available at http://nwql.usgs.gov/pubs/OFR/OFR-98-165.pdf.)

Glassmeyer, S.T., Furlong, E.T., Kolpin, D.W., Cahill, J.D., Zaugg, S.D., Werner, S.L., Meyer, M.T., and Kryak, D.D., 2005, Transport of chemical and microbial compounds from known wastewater discharges—Potential for use as indicators of human fecal contamination: Environmental Science and Technology, v. 39, p. 5,157–5,169.

Guy, H.P., 1969, Laboratory theory and methods for sediment analysis: U.S. Geological Survey Techniques of Water-Resources Investigations, book 5, chap. C1, 58 p. (Also available at http://pubs.usgs.gov/twri/twri5c1/html/pdf.html.)

Heberer, T., 2002, Occurrence, fate, and removal of pharmaceutical residues in the aquatic environment—A review of recent research data: Toxicology Letters, v. 131, p. 5–17.

Henny, C.J., Kaiser, J.L., Grove, R.A., Johnson, B.L., and Letcher, R.J., 2009, Polybrominated diphenyl ether flame retardants in eggs may reduce reproductive success of ospreys in Oregon and Washington, USA: Ecotoxicology, v. 18, p. 802–813.

Hirsch, R., Ternes, T., Haberer, K., and Kratz, K.L., 1999, Occurrence of antibiotics in the aquatic environment: The Science of the Total Environment, v. 225, p. 109–118.

Hoffman, G.L., Fishman, M.J., and Garbarino, J.R., 1996, Methods of analysis by the U.S. Geological Survey National Water Quality Laboratory—In-bottle acid digestion of whole-water samples: U.S. Geological Survey Open-File Report 96–225, 28 p.

Hoy, J.A., Haas, G.T., Hoy, R.D., and Hallock, P., 2011, Observations of brachygnathia superior in wild ruminants in Western Montana, USA: Wildlife Biology in Practice, v. 7, no. 2, 15 p.

Hu, Z., Shi, Y., and Cai, Y., 2011, Concentrations, distribution, and bioaccumulation of synthetic musks in the Haihe River of China: Chemosphere, v. 84, p. 1,630–1,635.

Hubbard, M.L., 2007, Analysis of the Oregon stakeholder drug take back public policy process to reduce pharmaceutical pollution in Oregon's water resources: Corvallis, Oregon State University Research Paper, 60 p., accessed March 2, 2012, at http://ir.library.oregonstate.edu/jspui/handle/1957/6192.

Independent Scientific Advisory Board, 2007, Human population impacts on Columbia River basin fish and wildlife: Portland, Oreg., Independent Scientific Advisory Board for the Northwest Power and Conservation Council, Columbia River Basin Indian Tribes, and National Marine Fisheries Service, ISAB 2007–3, 73 p., accessed March 2, 2012, at http://www.nwcouncil.org/library/isab/isab2007-3.pdf.

Ings, J.S., Servos, M.R., and Vijayan, M.M., 2011, Exposure to municipal wastewater effluent impacts stress performance in rainbow trout: Aquatic Toxicology, v. 103, p. 85–91.

International Flavors and Fragrances, Inc., 2007, Fragrance ingredients—Galaxolide 50 IPM: International Flavors and Fragrances, Inc., accessed March 2, 2012, at http://www.iff.com/Ingredients.nsf/0/445E7108D54E7AE6802569930038850E.

Jenkins, J.A., Goodbred, S.L., Sobiech, S.A., Olivier, H.M., Draugelis-Dale, R.O., and Alvarez, D.A., 2009, Effects of wastewater discharges on endocrine and reproductive function of western mosquitofish (*Gambusia* spp.) and implications for the threatened Santa Ana sucker (*Catostomus santaanae*): U.S. Geological Survey Open-File Report 2009–1097, 46 p. (Also available at http://pubs.usgs.gov/of/2009/1097/.)

Johnson, L.L., Collier, T.K., and Stein, J.E., 2002, An analysis in support of sediment quality thresholds for polycyclic aromatic hydrocarbons (PAHs) to protect estuarine fish: Aquatic Conservation—Marine and Freshwater Ecosystems, v. 12, p. 517–538.

Johnson, L.L., Ylitalo, G.M., Sloan, C.A., Anulacion, B.F., Kagley, A.N., Arkoosh, M.R., Lundrigan, T.A., Larson, K., Siipola, M., and Collier, T.K., 2007, Persistent organic pollutants in outmigrant juvenile chinook salmon from the Lower Columbia Estuary, USA: Science of the Total Environment, v. 374, p. 342–366.

Kaiser Family Foundation, 2010, Prescription drug trends—May 2010 update: Kaiser Family Foundation, 10 p., accessed March 2, 2012, at http://www.kff.org/rxdrugs/upload/3057-08.pdf.

Kidd, K.A., Blanchfield, P.J., Mills, K.H., Palace, V.P., Evans, R.E., Lazorchak, J.M., and Flick, R.W., 2007, Collapse of a fish population following exposure to a synthetic estrogen: Proceedings of the National Academy of Sciences of the United States of America, v. 104, no. 21, p. 8,897–8,901.

Kinney, C.A., Furlong, E.T., Kolpin, D.W., Burkhardt, M.R., Zaugg, S.D., Werner, S.L., Bossio, J.P., and Benotti, M.J., 2008, Bioaccumulation of pharmaceuticals and other anthropogenic waste indicators in earthworms from agricultural soil amended with biosolid or swine manure: Environmental Science and Technology, v. 42, no. 6, p. 1,863–1,870, accessed March 6, 2012, at http://pubs.acs.org/doi/abs/10.1021/es702304c.

Kinney, C.A., Furlong, E.T., Zaugg, S.D., Burkhardt, M.R., Werner, S.L., Cahill, J.D., and Jorgensen, G.R., 2006, Survey of organic wastewater contaminants in biosolids destined for land application: Environmental Science and Technology, v. 40, no. 23, p. 7,207–7,215.

Kolpin, D.W., Furlong, E.T., Meyer, M.T., Thurman, E.M., Zaugg, S.D., Barber, L.B., Buxton, H.T., 2002, Pharmaceuticals, hormones and other organic wastewater contaminants in U.S. streams, 1999–2000—A national reconnaissance: Environmental Science and Technology, v. 36, no. 6, p. 1,202–1,211, accessed March 1, 2012, at http://pubs.acs.org/doi/abs/10.1021/es011055j.

Kolpin, D.W., Skopec, M., Meyer, M.T., Furlong, E.T., and Zaugg, S.D., 2004, Urban contribution of pharmaceuticals and other organic wastewater contaminants to streams during differing flow conditions: Science of the Total Environment, v. 328, p. 119–130.

Koski, A.J., 2008, Control of annual grassy weeds in lawns: Colorado State University Extension Fact Sheet 3.101, 3 p., accessed May 7, 2012, at http://www.ext.colostate.edu/pubs/garden/03101.html.

Kümmerer, K., 2004, Resistance in the environment: Journal of Antimicrobial Chemotherapy, v. 54, p. 311–320.

Laetz, C.A., Baldwin, D.H., Collier, T.K., Hebert, V., Stark, J.D., and Scholz, N.L., 2009, The synergistic toxicity of pesticide mixtures—Implications for risk assessment and the conservation of endangered Pacific salmon: Environmental Health Perspectives, v. 117, no. 3, p. 348–353.

Lam, M.W., Young, C.J., Brain, R.A., Johnson, D.J., Hanson, M.A., Wilson, C.J., Richards, S.M., Solomon, K.R., and Mabury, S.A., 2004, Aquatic persistence of eight pharmaceuticals in a microcosm study: Environmental Toxicology and Chemistry, v. 23, p. 1,431–1,440.

Lindberg, R.H., Björklund, K., Rendahl, P., Johansson, M.I., Tysklind, M., and Andersson, B.A.V., 2007, Environmental risk assessment of antibiotics in the Swedish environment with emphasis on sewage treatment plants: Water Research, v. 41, p. 613–619.

Lindley, C.E., Stewart, J.T., and Sandstrom, M.W., 1996, Determination of low concentrations of acetochlor in water by automated solid-phase extraction and gas chromatography with mass selective detection: Journal of AOAC International, v. 79, no. 4, p. 962–966.

Lower Columbia River Estuary Partnership, 2007, Lower Columbia River and estuary ecosystem monitoring—Water quality and salmon sampling report: Portland, Oreg., Lower Columbia River Estuary Partnership, accessed March 1, 2012, at http://www.lcrep.org/sites/default/files/pdfs/WaterSalmonReport.pdf.

Lubick, N., 2010, Drugs in the environment—Do pharmaceutical take-back programs make a difference?: Environmental Health Perspectives, v. 118, p. a210–a214, accessed March 2, 2012, at http://ehp03.niehs.nih.gov/article/fetchArticle.action?articleURI=info%3Adoi%2F10.1289%2Fehp.118-a210.

Lubliner, B., Redding, M., and Golding, S., 2008, Quality assurance project plan—Pharmaceuticals and personal care products in wastewater treatment systems: Washington State Department of Ecology, Report #08–03–112, 36 p., accessed March 21, 2012, at http://www.ecy.wa.gov/biblio/0803112.html.

Madsen, J.E., Sandstrom, M.W., and Zaugg, S.D., 2003, Methods of analysis by the U.S. Geological Survey National Water Quality Laboratory—A method supplement for the determination of fipronil and degradates in water by gas chromatography/mass spectrometry: U.S. Geological Survey Open-File Report 02–462, 11 p. (Also available at http://nwql.usgs.gov/Public/pubs/OFR02-462/OFR02-462.html.)

Morace, J.L., 2006, Water-quality data, Columbia River Estuary, 2004–05: U.S. Geological Survey Data Series 213, 18 p. (Also available at http://pubs.water.usgs.gov/ds213.)

National Pesticide Information Center, 2009, Fipronil general fact sheet: Corvallis, Oreg., National Pesticide Information Center Fact Sheet, 3 p., accessed March 2, 2012, at http://npic.orst.edu/factsheets/fipronil.pdf.

Nelson, J., Bishay, F., van Roodselaar, A., Ikonomou, M., and Law, F.C.P., 2007, The use of in vitro bioassays to quantify endocrine disrupting chemicals in municipal wastewater treatment plant effluents: Science of the Total Environment, v. 374, p. 80–90.

Nilsen, E.B., Rosenbauer, R.R., Furlong, E.T., Burkhardt, M.R., Werner, S.L., Greaser, L., and Noriega, M., 2007, Pharmaceuticals, personal care products and anthropogenic waste indicators detected in streambed sediments of the Lower Columbia River and selected tributaries, in 6th International Conference on Pharmaceuticals and Endocrine Disrupting Chemicals in Water: Costa Mesa, Calif., National Ground Water Association, Paper 4483, 15 p.

Oblinger Childress, C.J., Foreman, W.T., Connor, B.F., and Maloney, T.J., 1999, New reporting procedures based on long-term method detection levels and some considerations for interpretations of water-quality data provided by the U.S. Geological Survey National Water Quality Laboratory: U.S. Geological Survey Open-File Report 99–193, 19 p. (Also available at http://water.usgs.gov/owq/OFR_99-193/.)

Oregon Department of Environmental Quality, 2010a, Memorandum from Dick Pedersen, Director, to the Environmental Quality Commission, regarding Agenda item J, Rule adoption—Persistent pollutant trigger levels: Oregon Department of Environmental Quality, May 26, 2010, 89 p., accessed March 1, 2012, at http://www.deq.state.or.us/about/eqc/agendas/attachments/2010june/J-PersistentPollutantTriggerLevels.pdf.

Oregon Department of Environmental Quality, 2010b, Toxics standards rule—Table 33A, Effective aquatic life criteria for federal Clean Water Act: Oregon Administrative Rules, OAR 340–041–033, accessed March 1, 2012, at http://www.deq.state.or.us/wq/rules/div041/table33a.pdf.

Oregon Department of Environmental Quality, 2011, Biosolids—A beneficial resource: Oregon Department of Environmental Quality, Water Quality Division, Biosolids Program, 05–WQ–002, 1 p.

Oregon Department of Environmental Quality and U.S. Environmental Protection Agency, 2005, Portland Harbor joint source control strategy, Final–December 2005: Oregon Department of Environmental Quality and U.S. Environmental Protection Agency, 81 p., accessed March 1, 2012, at http://www.deq.state.or.us/lq/cu/nwr/PortlandHarbor/docs/JSCSFinal0512.pdf.

Pait, A.S., Warner, R.A., Hartwell, S.I., Nelson, J.O., Pacheco, P.A., and Mason, A.L., 2006, Human use pharmaceuticals in the estuarine environment—A survey of the Chesapeake Bay, Biscayne Bay and Gulf of the Farallones: Silver Spring, Md., National Oceanic and Atmospheric Administration National Centers for Coastal Ocean Science, Center for Coastal Monitoring and Assessment, 21 p.

Phillips, P.J., and Chalmers, A.T., 2009, Wastewater effluent, combined sewer overflows, and other sources of organic compounds to Lake Champlain: Journal of the American Water Works Association, v. 45, no. 1, p. 45–57. (Also available at http://onlinelibrary.wiley.com/doi/10.1111/j.1752-1688.2008.00288.x/abstract.)

Portland Bureau of Environmental Services, 2010, Controlling combined sewer overflows: City of Portland, accessed March 1, 2012, at http://www.portlandonline.com/bes/index.cfm?c=31030.

PubMed Health, 2008, Trimethoprim: U.S. National Library of Health and National Institutes of Health, accessed March 12, 2012, at http://www.ncbi.nlm.nih.gov/pubmedhealth/PMH0000813.

PubMed Health, 2010, Diphenhydramine: U.S. National Library of Health and National Institutes of Health, accessed March 12, 2012, at http://www.ncbi.nlm.nih.gov/pubmedhealth/PMH0000704.

Rastall, A.C., Neziri, A., Vukonvic, Z., Jung, C., Mijovic, S., Hollert, H., Nikcevic, S., and Erdinger, L., 2004, The identification of readily bioavailable pollutants in Lake Shkodra/Skadar using semipermeable membrane devices (SPMDs), bioassays and chemical analysis: Environmental Science and Pollution Research, v. 11, p. 240–253.

Rounds, S.A., Doyle, M.C., Edwards, P.M., and Furlong, E.T., 2009, Reconnaissance of pharmaceutical chemicals in urban streams of the Tualatin River basin, Oregon, 2002: U.S. Geological Survey Scientific Investigations Report 2009–5119, 22 p. (Also available at http://pubs.usgs.gov/sir/2009/5119/index.html.)

Routledge, E.J., and Sumpter, J.P., 1996, Estrogenic activity of surfactants and some of their degradation products assessed using a recombinant yeast screen: Environmental Toxicology and Chemistry, v. 15, p. 241–248.

Sandahl, J.F., Baldwin, D.H., Jenkins, J.J., and Scholz, N.L., 2007, A sensory system at the interface between urban stormwater runoff and salmon survival: Environmental Science and Technology, v. 41, p. 2,998–3,004.

Sandstrom, M.W., Stroppel, M.E., Foreman, W.T., and Schroeder, M.P., 2001, Methods of analysis by the U.S. Geological Survey National Water Quality Laboratory—Determination of moderate-use pesticides and selected degradates in water by C-18 solid-phase extraction and gas chromatography/mass spectrometry: U.S. Geological Survey Water-Resources Investigations Report 01–4098, 70 p. (Also available at http://nwql.usgs.gov/Public/pubs/WRIR01-4098.html.)

Sansalone, J.J., Koran, J.M., Smithson, J.A., and Buchberger, S.G., 1998, Physical characteristics of urban roadway solids transported during rain events: Journal of Environmental Engineering, v. 124, p. 427–440.

Scholz, N.L., Truelove, N.K., Labeniz, J.S., Baldwin, D.H., and Collier, T.K., 2006, Dose-additive inhibition of chinook salmon acetylcholinesterase activity by mixtures of organophosphate and carbamates insecticides: Environmental Toxicology and Chemistry, v. 25, p. 1,200–1,207.

Schreurs, R.H.M.M., Legler, J., Artola-Garicano, E., Sinnige, T.L., Lanser, P.H., Seinen, W., and Van Der Burg, B., 2004, *In vitro* and *in vivo* antiestrogenic effects of polycyclic musks in zebrafish: Environmental Science and Technology, v. 38, no. 4, p. 997–1,002.

Schreurs, R.H.M.M., Sonneveld, E., Jansen, J.H., Seinen, W., and Van Der Burg, B., 2005, Interaction of polycyclic musks and UV filters with the estrogen receptor (ER), androgen receptor (AR), and progesterone receptor (PR) in reporter gene bioassays: Toxicological Sciences, v. 83, p. 264–272.

Sierra Club, 2005, Nonylphenol ethoxylates—A safer alternative exists to this toxic cleaning agent: Sierra Club, 11 p., accessed March 1, 2012, at http://www.jcaa.org/news/references/Sierra%20Club%20a%20safer%20alternative%20nonylphenol_ethoxylates3%5B1%5D.pdf.

Smital, T., 2008, Acute and chronic effects of emerging contaminants, *in* Barceló, D., and Petrovic, M., *eds.,* Emerging contaminants from industrial and municipal waste—Occurrence, analysis and effects: Berlin, Springer-Verlag, p. 105–142.

Sprague, L.A., and Battaglin, W.A., 2005, Wastewater chemicals in Colorado's streams and ground water: U.S. Geological Survey Fact Sheet 2004–3127, 4 p. (Also available at http://pubs.usgs.gov/fs/2004/3127/.)

Ternes, T.A., 1998, Occurrence of drugs in German sewage treatment plants and rivers: Water Research, v. 32, p. 3,245–3,260.

Tilton, S.C., Foran, C.M., and Benson, W.H., 2003, Effects of cadmium on the reproductive axis of Japanese medaka (*Oryzias latipes*)—Comparative Biochemistry and Physiology–Part C: Toxicology and Pharmacology, v. 136, no. 3, p. 265–76.

U.S. Census Bureau, 2010, American factfinder fact sheets: U.S. Census Bureau database, accessed March 6, 2012, at http://factfinder2.census.gov.

U.S. Environmental Protection Agency, 1999, Method 1664, revision A—N-hexane extractable material (HEM; oil and grease) and silica gel treated N-hexane extractable material (SGT-HEM—nonpolar material) by extraction and gravimetry: U.S. Environmental Protection Agency Office of Water, EPA 821/R–98–002, 28 p.

U.S. Environmental Protection Agency, 2002, Method 1631—Measurement of mercury in water— Revision: U.S. Environmental Protection Agency, Office of Water report, accessed March 1, 2012, at http://water.epa.gov/ scitech/methods/cwa/metals/mercury/upload/2007_07_10_ methods_method_mercury_1631.pdf.

U.S. Environmental Protection Agency, 2006, Aquatic life criteria for nonylphenol—Final aquatic life ambient water quality criteria—Nonylphenol: U.S. Environmental Protection Agency Fact Sheet EPA–822–F–05–003, accessed March 1, 2012 at http://water.epa.gov/scitech/ swguidance/standards/criteria/aqlife/pollutants/nonylphenol/ nonylphenol-fs.cfm.

U.S. Environmental Protection Agency, 2008, Chromated copper arsenate (CCA)—Propiconazole, an alternative to CCA: U.S. Environmental Protection Agency web page, accessed March 1, 2012, at http://www.epa.gov/oppad001/ reregistration/cca/propiconazole htm.

U.S. Environmental Protection Agency, 2009a, Columbia River basin—State of the river report for toxics: U.S. Environmental Protection Agency, EPA 910–R–08–004, accessed March 1, 2012, at http://yosemite.epa.gov/r10/ ECOCOMM.NSF/Columbia/SoRR/.

U.S. Environmental Protection Agency, 2009b, National recommended water quality criteria: U.S. Environmental Protection Agency, Offices of Water and Science and Technology, accessed March 1, 2012, at http://www.epa. gov/ost/criteria/wqctable/.

U.S. Environmental Protection Agency, 2010, Region 10 cleanup—Portland Harbor: U.S. Environmental Protection Agency, accessed March 1, 2012, at http://yosemite.epa. gov/r10/cleanup nsf/cff418266f1ddba08825777d007dffb4! OpenView.

U.S. Geological Survey, variously dated, National field manual for the collection of water-quality data: U.S. Geological Survey Techniques of Water-Resources Investigations, book 9, chaps. A1–A9, 2 v., variously paged, accessed March 14, 2012, at http://pubs.water.usgs.gov/twri9A/.

Vajda, A.M., Barber, L.B., Gray, J.L., Lopez, E.M., Woodling, J.D., and Norris, D.O., 2008, Reproductive disruption in fish downstream from an estrogenic wastewater effluent: Environmental Science and Technology, v. 42, p. 3,407–3,414.

VanMetre, P.C., Mahler, B.J., and Wilson, J.T., 2009, PAHs underfoot—Contaminated dust from coal-tar sealcoated pavement is widespread in the United States: Environmental Science and Technology, v. 43, no. 1, p. 20–25.

Vermeirssen, E.L.M., Burki, R., Joris, C., Peter, A., and Segner, H., 2005, Characterization of the estrogenicity of Swiss midland rivers using a recombinant yeast bioassay and plasma vitellogenin concentrations in feral male brown trout: Environmental Toxicology and Chemistry, v. 24, no. 9, p. 2,226–2,233.

Washington State Department of Ecology, 2003, Water quality standards for surface waters of the State of Washington: Washington State Legislature, Washington Administrative Code, chap. 173–201A WAC, 101 p., accessed March 6, 2012, at http://apps.leg.wa.gov/wac/default.aspx?cite=173- 201A&full=true#173-201A-240.

Washington State Department of Health, 2007, Wenatchee River fish consumption advice: Washington State Department of Health, DOH publication #334–131, June 2007, 2 p., accessed March 14, 2012, at http://www. doh.wa.gov/ehp/oehas/pubs/wenatcheeriver.pdf.

Williams, R.T., ed., 2005, Human pharmaceuticals—Assessing the impacts on aquatic ecosystems: Pensacola, Fla., Society of Environmental Toxicology and Chemistry Press, 368 p.

Wilson, B.A., Smith, V.H., Denoyelles, F., Larive, C.K., 2003, Effects of three pharmaceutical and personal care products on natural freshwater algal assemblages: Environmental Science and Technology, v. 37, no. 9, p. 1,713–1,719.

Yoqui, G.T., and Sericano, J.L., 2009, Polybrominated diphenyl ether flame retardants in the U.S. marine environment—A review: Environment International, v. 35, no., 3, p. 655–666, accessed March 23, 2012, at http://www. ncbi nlm nih.gov/pubmed/19100622.

Zaugg, S.D., Sandstrom, M.W., Smith, S.G., and Fehlberg, K.M., 1995, Methods of analysis by the U.S. Geological Survey National Water Quality Laboratory—Determination of pesticides in water by C-18 solid-phase extraction and capillary-column gas chromatography/mass spectrometry with selected-ion monitoring: U.S. Geological Survey Open-File Report 95–181, 60 p. (Also available at http://nwql.usgs.gov/Public/pubs/OFR95-181/OFR95-181.html.)

Zaugg, S.D., Smith, S.G., and Schroeder, M.P., 2006, Methods of analysis by the U.S. Geological Survey National Water Quality Laboratory—Determination of wastewater compounds in whole water by continuous liquid-liquid extraction and capillary-column gas chromatography/mass spectrometry: U.S. Geological Survey Techniques and Methods, book 5, chap. B4, 30 p. (Also available at http://pubs.usgs.gov/tm/2006/05B04/.)

The Columbia River at The Dalles, Oregon, December 2008.

Aerated stabilization basin at City of St. Helens Wastewater-Treatment Plant, Oregon, December 2009.

Appendix A. – Methods, Reporting Limits, and Analyte Information

Table A1. Reporting limits and possible uses or sources of halogenated compounds analyzed in solids filtered from wastewater-treatment-plant effluent or stormwater runoff, Columbia River Basin, Washington and Oregon, 2008–10.

[Shading indicates a detection in wastewater-treatment-plant effluent in this study; bold type indicates a detection in stormwater runoff in this study. **CAS No.:** Chemical Abstract Service Registry Number® is a Registered Trademark of the American Chemical Society. CAS recommends the verification of the CASRNs through CAS Client Services[SM]. **Reporting limit:** Reported in nanograms; divide the reporting limit by the volume of water filtered for each sample to get the reporting limit in nanograms per liter. **Possible compound uses or sources:** From Steve Zaugg, U.S. Geological Survey, written commun., December 2008. **Abbreviation:** NA, not available]

Analyte	Parameter code	CAS No.	Reporting limit	Possible compound uses or sources
Polybrominated diphenyl ethers (PBDEs) and other flame retardants				
Bis(hexachlorocyclopentadieno) cyclooctane [Dechlorane Plus]	65220	13560-89-9	1	Chlorinated flame retardant
1,2-Bis(2,4,6-tribromophenoxy) ethane [Firemaster 680]	64868	37853-59-1	0.1	Brominated flame retardant
Pentabromotoluene	64867	87-83-2	1	Brominated flame retardant
2,2',4,4'-Tetrabromodiphenylether (PBDE-47)	63166	5436-43-1	0.2	Textile and electronic flame retardant
2,3',4,4'-Tetrabromodiphenyl ether (PBDE-66)	64852	189084-61-5	0.1	Textile and electronic flame retardant
2,3',4',6-Tetrabromodiphenyl ether (PBDE-71)	64853	189084-62-6	0.1	Textile and electronic flame retardant
2,2',3,4,4'-Pentabromodiphenyl ether (PBDE-85)	64854	182346-21-0	0.1	Textile and electronic flame retardant
2,2',4,4',5-Pentabromodiphenyl ether (PBDE-99)	64855	60348-60-9	0.2	Textile and electronic flame retardant
2,2',4,4',6-Pentabromodiphenyl ether (PBDE-100)	64856	189084-64-8	0.1	Textile and electronic flame retardant
2,2',3,4,4',5'-Hexabromodiphenyl ether (PBDE-138)	64857	182677-30-1	0.1	Textile and electronic flame retardant
2,2',4,4',5,5'-Hexabromodiphenylether (PBDE-153)	64858	68631-49-2	0.1	Textile and electronic flame retardant
2,2',4,4',5,6'-Hexabromodiphenyl ether (PBDE-154)	64859	207122-15-4	0.1	Textile and electronic flame retardant
2,2',3,4,4',5',6-Heptabromodiphenyl ether (PBDE-183)	64860	207122-16-5	0.1	Textile and electronic flame retardant
Polychlorinated biphenyls (PCBs)				
PCB-49	64725	41464-40-8	2	PCB congener
PCB-52	64726	35693-99-3	1	PCB congener
PCB-70	64727	32598-11-1	2	PCB congener
PCB-101	64729	37680-73-2	1	PCB congener
PCB-110	64730	38380-03-9	1	PCB congener
PCB-118	64731	31508-00-6	0.1	PCB congener
PCB-138	64732	35065-28-2	0.1	PCB congener
PCB-146	64733	51908-16-8	0.1	PCB congener
PCB-149	64734	38380-04-0	1	PCB congener
PCB-151	64735	52663-63-5	0.1	PCB congener
PCB-170	64736	35065-30-6	0.1	PCB congener
PCB-174	64737	38411-25-5	0.1	PCB congener
PCB-177	64738	52663-70-4	0.1	PCB congener
PCB-180	64739	35065-29-3	0.1	PCB congener
PCB-183	64740	52663-69-1	0.1	PCB congener
PCB-187	64741	52663-68-0	0.1	PCB congener
PCB-194	64742	35694-08-7	0.1	PCB congener
PCB-206	64743	40186-72-9	0.1	PCB congener

Table A1. Reporting limits and possible uses or sources of halogenated compounds analyzed in solids filtered from wastewater-treatment-plant effluent or stormwater runoff, Columbia River Basin, Washington and Oregon, 2008–10.—Continued

[Shading indicates a detection in wastewater-treatment-plant effluent in this study; bold type indicates a detection in stormwater runoff in this study. **CAS No.:** Chemical Abstract Service Registry Number® is a Registered Trademark of the American Chemical Society. CAS recommends the verification of the CASRNs through CAS Client Services[SM]. **Reporting limit:** Reported in nanograms; divide the reporting limit by the volume of water filtered for each sample to get the reporting limit in nanograms per liter. **Possible compound uses or sources:** From Steve Zaugg, U.S. Geological Survey, written commun , December 2008. **Abbreviation:** NA, not available]

Analyte	Parameter code	CAS No.	Reporting limit	Possible compound uses or sources
Herbicides and insecticides				
Benfluralin	63265	1861-40-1	0.2	Dinitroaniline herbicide
cis-Chlordane	63271	5103-71-9	0.2	Organochlorine insecticide
trans-Chlordane	63272	5103-74-2	0.2	Organochlorine insecticide
Chlorpyrifos	63273	2921-88-2	0.2	Organophosphate insecticide
Cyfluthrin	63279	68359-37-5	0.2	Insecticide
lambda-Cyhalothrin	63280	91465-08-6	0.2	Insecticide
Dacthal (DCPA)	63282	1861-32-1	0.2	Phenoxyacid herbicide
p,p'-dichlorodiphenyldichloroethane (DDD)	63346	72-54-8	2	Legacy pesticide
p,p'-dichlorodiphenyldichloroethylene (DDE)	63347	72-55-9	1	Legacy pesticide
p,p'-dichlorodiphenyltrichloroethane (DDT)	63345	50-29-3	4	Legacy pesticide
Desulfinylfipronil	63316	NA	0.1	Fipronil degradate
Dieldrin	63289	60-57-1	0.1	Organochlorine insecticide
alpha-Endosulfan	63259	959-98-8	0.2	Organochlorine insecticide
Fipronil	63313	120068-37-3	0.1	Insecticide
Fipronil sulfide	63314	120067-83-6	0.1	Fipronil degradate
cis-Nonachlor	63338	5103-73-1	0.1	Insecticide
trans-Nonachlor	63339	39765-80-5	0.1	Insecticide
Oxychlordane	64866	27304-13-8	1	Insecticide
Oxyfluorfen	63341	42874-03-3	4	Herbicide
Pendimethalin	63353	40487-42-1	1	Herbicide
Pentachloroanisole (Chloridazon)	64119	1825-21-4	0.1	Herbicide
Pentachloronitrobenzene	63650	82-68-8	0.1	Organochlorine herbicide
Tefluthrin	63377	79538-32-2	0.5	Insecticide
Trifluralin	63390	1582-09-8	0.2	Dinitroaniline herbicide
Other compounds				
Hexachlorobenzene (HCB)	63631	118-74-1	0.1	Organochlorine fungicide
Methoxy triclosan	63639	1000766	6	Triclosan degradate
Octachlorostyrene	65217	29082-74-4	1	Combustion by-product
Tetradifon	63665	116290	0.2	Acaricide
Triclosan	63232	3380-34-5	4	Anti-bacterial agent

Table A2. Reporting limits and possible uses or sources of anthropogenic organic compounds analyzed in unfiltered wastewater-treatment-plant effluent, Columbia River Basin, Washington and Oregon, 2008–09.

[Shading indicates a detection in wastewater-treatment-plant effluent in this study. **CAS No.:** Chemical Abstract Service Registry Number® is a Registered Trademark of the American Chemical Society. CAS recommends the verification of the CASRNs through CAS Client Services[SM]. **Method detection limit** and **Reporting limit:** Values are in micrograms per liter. **EDP** (endocrine-disrupting potential) and **Possible compound uses or sources:** From Zaugg and others (2006). **Abbreviations:** K, known; S, suspected; >, greater than; –, no data; NA, not available]

Analyte	Parameter code	CAS No.	Method detection limit	Reporting limit	EDP	Possible compound uses or sources
			Detergent metabolites			
4-Cumylphenol	62808	599-64-4	0.13	0.2	K	Nonionic detergent metabolite
para-Nonylphenol (total)	62829	84852-15-3	1.2	1.6	K	Nonionic detergent metabolite
4-Nonylphenol monoethoxylate (sum of all isomers) [NP1EO]	61704	NA	1.35	1.6	K	Nonionic detergent metabolite
4-Nonylphenol diethoxylate (sum of all isomers) [NP2EO]	61703	NA	1.2	3.2	K	Nonionic detergent metabolite
4-n-Octylphenol	62809	1806-26-4	0.11	0.2	K	Nonionic detergent metabolite
4-tert-Octylphenol	62810	140-66-9	0.11	0.4	K	Nonionic detergent metabolite
4-tert-Octylphenol diethoxylate (OP2EO)	62486	NA	0.05	10.32	K	Nonionic detergent metabolite
4-tert-Octylphenol monoethoxylate (OP1EO)	62485	NA	0.5	1	K	Nonionic detergent metabolite
			Flame retardants			
2,2',4,4'-Tetrabromodiphenyl ether (PBDE-47)	63147	5436-43-1	0.11	0.3	–	Widely used brominated flame retardant
Tri(2-butoxyethyl)phosphate	62830	78-51-3	0.05	0.2	–	Flame retardant
Tri(2-chloroethyl)phosphate	62831	115-96-8	0.08	0.2	S	Plasticizer, flame retardant
Tri(dichlorisopropyl)phosphate	61707	13674-87-8	0.05	0.2	S	Flame retardant
Tributyl phosphate	62832	126-73-8	0.11	0.2	–	Antifoaming agent, flame retardant
			Miscellaneous - antioxidants, solvents, or multiple uses			
Anthraquinone	62813	84-65-1	0.08	0.2	–	Manufacturing of dye/textures, seed treatment, bird repellant
Bromoform	32104	75-25-2	0.03	0.2	–	Wastewater ozination byproduct, military/explosives
3-tert-Butyl-4-hydroxyanisole (BHA)	61702	25013-16-5	0.16	0.2	K	Antioxidant, general preservative
Caffeine	81436	58-08-2	0.06	0.2	–	Beverages, diuretic, very mobile/biodegradable
Carbazole	77571	86-74-8	0.12	0.2	–	Insecticide, manufacturing of dyes, explosives, and lubricants
Cotinine	61945	486-56-6	0.29	0.8	–	Primary nicotine metabolite
para-Cresol	77146	106-44-5	0.08	0.2	S	Wood preservative
Isophorone	34408	78-59-1	0.08	0.2	–	Solvent for lacquer, plastic, oil, silicon, resin
Isopropylbenzene (cumene)	77223	98-82-8	0.02	0.2	–	Manufacturing of phenol/acetone, fuels and paint thinner
d-Limonene	62819	5989-27-5	0.02	0.2	–	Fungicide, antimicrobial, antiviral, fragrance in aerosols
5-Methyl-1H-benzotriazole	61944	136-85-6	0.35	1.6	–	Antioxidant in antifreeze and deicers
Pentachlorophenol	39032	87-86-5	0.33	0.8	S	Herbicide, fungicide, wood preservative, termite control
Tetrachloroethylene	34475	127-18-4	0.22	0.4	–	Solvent, degreaser, veterinary antihelmintic

Table A2. Reporting limits and possible uses or sources of anthropogenic organic compounds analyzed in unfiltered wastewater-treatment-plant effluent, Columbia River Basin, Washington and Oregon, 2008–09.—Continued

[Shading indicates a detection in wastewater-treatment-plant effluent in this study. **CAS No.:** Chemical Abstract Service Registry Number® is a Registered Trademark of the American Chemical Society. CAS recommends the verification of the CASRNs through CAS Client Services^{SM}. **Method detection limit** and **Reporting limit:** Values are in micrograms per liter. **EDP** (endocrine-disrupting potential) and **Possible compound uses or sources:** From Zaugg and others (2006). **Abbreviations:** K, known; S, suspected; >, greater than; –, no data; NA, not available]

Analyte	Parameter code	CAS No.	Method detection limit	Reporting limit	EDP	Possible compound uses or sources
Personal care products						
Acetophenone	62811	98-86-2	0.07	0.4	–	Fragrance in detergent and tobacco, flavor in beverages
Benzophenone	62814	119-61-9	0.1	0.2	S	Fixative for perfumes and soaps
Camphor	62817	76-22-2	0.09	0.2	–	Flavor, odorant, ointments
1,4-Dichlorobenzene	34571	106-46-7	0.03	0.2	S	Moth repellent, fumigant, deodorant
Galaxolide (hexahydrohexamethylcyclopentabenzopyran, HHCB)	62823	1222-05-5	0.11	0.2	–	Musk fragrance (widespread usage) persistent in ground water
Indole	62824	120-72-9	0.08	0.2	–	Pesticide inert ingredient, fragrance in coffee
Isoborneol	62825	124-76-5	0.05	0.2	–	Fragrance in perfumery, in disinfectants
Isoquinoline	62826	119-65-3	0.09	0.2	–	Flavors and fragrances
Menthol	62827	89-78-1	0.05	0.2	–	Cigarettes, cough drops, liniment, mouthwash
3-Methyl-1H-indole (skatol)	62807	83-34-1	0.07	0.2	–	Fragrance, stench in feces and coal tar
Methyl salicylate	62828	119-36-8	0.07	0.2	–	Liniment, food, beverage, ultraviolet-absorbing lotion
Phenol	34694	108-95-2	0.07	0.2	–	Disinfectant, manufacturing of several products, leachate
Tonalide (acetyl-hexamethyl-tetrahydronaphthalene, AHTN)	62812	21145-77-7	0.11	0.2	–	Musk fragrance (widespread usage) persistent in ground water
Triclosan	61708	3380-34-5	0.09	0.2	S	Disinfectant, antimicrobial (concern for acquired microbial resistance)
Triethyl citrate (ethyl citrate)	62833	77-93-0	0.07	0.2	–	Cosmetics, pharmaceuticals
Pesticides						
Atrazine	39630	1912-24-9	0.08	0.2	K	Selective triazine herbicide
Bromacil	30234	314-40-9	0.1	0.8	S	General-use herbicide, >80 percent noncrop usage on grass/brush
Carbaryl	39750	63-25-2	0.13	0.2	K	Insecticide, crop and garden uses, low persistence
Chlorpyrifos	38932	2921-88-2	0.12	0.2	K	Insecticide, domestic pest/termite control (domestic use restricted as of 2001)
Diazinon	39570	333-41-5	0.11	0.2	K	Insecticide, >40 percent nonagricultural usage, ants, flies
3,4-Dichlorophenyl isocyanate	63145	102-36-3	0.06	1.6	–	Degradate of diuron, a noncrop herbicide
Dichlorvos	30218	62-73-7	0.11	0.2	S	Insecticide, pet collars; Naled or Trichlorfon degradate
Metalaxyl	4254	57837-19-1	0.13	0.2	–	Herbicide, fungicide, general-use pesticide, mildew, blight, pathogens, golf/turf
Metolachlor	82612	51218-45-2	0.12	0.2	–	Herbicide, general-use pesticide, indicator of agricultural drainage
N,N-diethyl-meta-toluamide (Deet)	61947	134-62-3	0.12	0.2	–	Insecticide, urban uses, mosquito repellent
Prometon	39056	1610-18-0	0.08	0.2	–	Herbicide (noncrop only), applied prior to blacktop

Table A2. Reporting limits and possible uses or sources of anthropogenic organic compounds analyzed in unfiltered wastewater-treatment-plant effluent, Columbia River Basin, Washington and Oregon, 2008–09.—Continued

[Shading indicates a detection in wastewater-treatment-plant effluent in this study. **CAS No.:** Chemical Abstract Service Registry Number® is a Registered Trademark of the American Chemical Society. CAS recommends the verification of the CASRNs through CAS Client Services[SM]. **Method detection limit** and **Reporting limit:** Values are in micrograms per liter. **EDP** (endocrine-disrupting potential) and **Possible compound uses or sources:** From Zaugg and others (2006). **Abbreviations:** K, known; S, suspected; >, greater than; —, no data; NA, not available]

Analyte	Parameter code	CAS No.	Method detection limit	Reporting limit	EDP	Possible compound uses or sources
Plasticizers						
Bisphenol A	62816	80-05-7	0.22	0.4	K	Manufacturing of polycarbonate resins, antioxidant, flame retardant
Diethyl phthalate (DEP)	34336	84-66-2	0.1	0.2	K	Plasticizer for polymers and resins
bis-(2-Ethylhexyl) phthalate (DEHP)	39100	117-81-7	0.85	2	K	Plasticizer for polymers and resins, pesticide inert
Triphenyl phosphate	62834	115-86-6	0.1	0.2	–	Plasticizer, resin, wax, finish, roofing paper, flame retardant
Polycyclic aromatic hydrocarbons (PAHs)						
Anthracene	34220	120-12-7	0.08	0.2	–	Wood preservative, component of tar, diesel, or crude oil, combustion product
Benzo(a)pyrene	34247	50-32-8	0.06	0.2	K	Regulated polycyclic aromatic hydrocarbon (PAH), used in cancer research, combustion product
2,6-Dimethylnaphthalene	62805	581-42-0	0.05	0.2	–	Present in diesel/kerosene (trace in gasoline)
Fluoranthene	34376	206-44-0	0.08	0.2	–	Component of coal tar, asphalt (only traces in gasoline or diesel fuel), combustion product
1-Methylnaphthalene	81696	90-12-0	0.03	0.2	–	2–5 percent of gasoline, diesel fuel, or crude oil
2-Methylnaphthalene	30194	91-57-6	0.03	0.2	–	2–5 percent of gasoline, diesel fuel, or crude oil
Naphthalene	34696	91-20-3	0.03	0.2	–	Fumigant, moth repellent, major component (about 10 percent) of gasoline
Phenanthrene	34461	85-01-8	0.07	0.2	–	Manufacturing of explosives, component of tar, diesel fuel, or crude oil, combustion product
Pyrene	34469	129-00-0	0.08	0.2	–	Component of coal tar, asphalt (only traces in gasoline or diesel fuel), combustion product
Steroids						
Cholesterol	62818	57-88-5	0.3	1.6	–	Often a fecal indicator, also a plant sterol
3-beta-Coprostanol	62806	360-68-9	0.38	1.6	–	Carnivore fecal indicator
beta-Sitosterol	62815	83-46-5	0.11	1.6	–	Plant sterol
beta-Stigmastanol	61948	19466-47-8	0.22	1.7	–	Plant sterol

[1]The reporting limit for 4-tert-Octylphenol diethoxylate (OP2EO) changed to 0.5 micrograms per liter in 2009.

Table A3. Reporting limits and intended uses of pharmaceuticals analyzed in filtered wastewater-treatment-plant effluent, Columbia River Basin, Washington and Oregon, 2008–09.

[Shading indicates a detection in wastewater-treatment-plant effluent in this study. **CAS No.:** Chemical Abstract Service Registry Number® is a Registered Trademark of the American Chemical Society. CAS recommends the verification of the CASRNs through CAS Client Services[SM]. **Method detection limit** and **Reporting limit:** Values are in micrograms per liter. **Common name** and **Intended use:** From Rounds and others, 2009; **Symbol:** –, not applicable]

Analyte	CAS No.	Method detection limit		Reporting limit		Common name	Intended use
		2008	2009	2008	2009		
Acetaminophen	103-90-2	0.04	0.06	0.08	0.12	Tylenol®	Analgesic
Albuterol	18559-94-9	0.03	0.04	0.06	0.08	Ventolin®, Airomir™ MDI	Bronchodilator (for asthma)
Caffeine	58-08-2	0.1	0.03	0.2	0.06	NoDoz®	Stimulant
Carbamazepine	298-46-4	0.02	0.03	0.04	0.06	Epitol®, Tegretol®	Anticonvulsant; antimanic (mood stabilizer)
Codeine	76-57-3	0.02	0.023	0.04	0.046	Robitussin® AC	Opioid narcotic; cough suppressant
Cotinine	486-56-6	0.01	0.019	0.026	0.038	–	Metabolite of nicotine
Dehydronifedipine	67035-22-7	0.04	0.04	0.08	0.08	–	Metabolite and photodegradation product of nifedipine (arterial dilator; antihypertensive)
Diltiazem	42399-41-7	0.04	0.03	0.08	0.06	Cardizem®	Antianginal; antiarrhythmic; antihypertensive
1,7-Dimethylxanthine	611-59-6	0.06	0.05	0.12	0.1	–	Metabolite of caffeine
Diphenhydramine	147-24-0	0.02	0.018	0.04	0.036	Benadryl®, Allerdryl	Antihistamine; sedative
Sulfamethoxazole	723-46-6	0.08	0.08	0.16	0.16	Bactrim™, Septra® (component)	Antibiotic; antibacterial
Thiabendazole	148-79-8	0.03	0.03	0.06	0.06	Arbotect®, Mertect®	Systemic antifungal; livestock antiparisitic
Trimethoprim	738-70-5	0.01	0.017	0.02	0.034	Bactrim, Septra (component)	Antibiotic; antibacterial
Warfarin	81-81-2	0.05	0.04	0.1	0.08	Coumadin®	Anticoagulant

Table A4. Reporting limits, uses, and classes of currently used pesticides and degradates analyzed in filtered wastewater-treatment-plant effluent or stormwater runoff, Columbia River Basin, Washington and Oregon, 2009–10.

[Shading indicates a detection in wastewater-treatment-plant effluent; bold type indicates a detection in stormwater runoff in this study. **CAS No.:** Chemical Abstract Service Registry Number® is a Registered Trademark of the American Chemical Society. CAS recommends the verification of the CASRNs through CAS Client Services^SM. **Method detection limit** and **Reporting limit:** Values are in micrograms per liter. **Abbreviations:** NA, not available; D, degradate; Def, defoliant; F, fumigant; Fun, fungicide; H, herbicide; I, insecticide; N, nematocide; –, not applicable]

Analyte	Parameter code	CAS No.	Method detection limit		Reporting limit		Use	Class	Parent compound
			February–September 2009	October 2009–June 2010	February–September 2009	October 2009–June 2010			
Fungicides									
3,5-Dichloroaniline	61627	626-43-7	0.002	0.0015	0.004	0.003	D	Amide	Iprodione
Iprodione	61593	36734-19-7	0.007	0.007	0.014	0.014	Fun	Amide	–
Metalaxyl	61596	57837-19-1	0.003	0.003	0.007	0.007	Fun	Acylalanine	–
Myclobutanil	61599	88671-89-0	0.005	0.005	0.01	0.01	Fun	Triazole	–
cis-Propiconazole	79846	60207-90-1	0.003	0.003	0.006	0.006	Fun	Triazole	–
trans-Propiconazole	79847	60207-90-1	0.01	0.01	0.02	0.02	Fun	Triazole	–
Tebuconazole	62852	107534-96-3	0.01	0.01	0.02	0.02	Fun	Triazole	–
Herbicides and degradates									
Acetochlor	49260	34256-82-1	0.005	0.005	0.01	0.01	H	Chloroacetamide	–
Alachlor	46342	15972-60-8	0.004	0.004	0.008	0.008	H	Chloroacetamide	–
Atrazine	39632	1912-24-9	0.004	0.004	0.007	0.007	H	Triazine	–
Benfluralin	82673	1861-40-1	0.007	0.007	0.014	0.014	H	Dinitroaniline	–
2-Chloro-2,6-diethylacetanilide	61618	6967-29-9	0.005	0.005	0.01	0.01	D	Chloroacetamide	Alachlor
2-Chloro-4-isopropylamino-6-amino-s-triazine (CIAT)	04040	6190-65-4	0.007	0.007	0.014	0.014	D	Triazine	Atrazine
4-Chloro-2-methylphenol	61633	1570-64-5	0.0025	0.002	0.005	0.0032	D	Chlorophenoxy acid	MCPA, MCPB
Cyanazine	04041	21725-46-2	0.02	0.011	0.04	0.022	H	Triazine	–
DCPA (Dacthal)	82682	1861-32-1	0.003	0.0038	0.006	0.0076	H	Organochlorine	–
3,4-Dichloroaniline	61625	95-76-1	0.002	0.0021	0.004	0.0042	D	Phenyl urea, Chloroacetamide	Diuron, Propanil, Linuron, Neburon
2,6-Diethylaniline	82660	579-66-8	0.003	0.003	0.006	0.006	D	Chloroacetamide	Alachlor
S-Ethyl dipropyl thiocarbamate (EPTC)	82668	759-94-4	0.001	0.001	0.002	0.002	H	Thiocarbamate	–
2-Ethyl-6-methylaniline	61620	24549-06-2	0.0049	0.0049	0.010	0.0098	D	Chloroacetamide	Metolachlor
Hexazinone	04025	51235-04-2	0.004	0.004	0.008	0.008	H	Triazine	–
Metolachlor	39415	51218-45-2	0.007	0.007	0.014	0.014	H	Chloroacetamide	–
Metribuzin	82630	21087-64-9	0.008	0.006	0.016	0.012	H	Triazine	–
Molinate	82671	2212-67-1	0.001	0.0013	0.002	0.0026	H	Thiocarbamate	–
Oxyfluorfen	61600	42874-03-3	0.003	0.005	0.006	0.01	H	Diphenyl ether	–
Pendimethalin	82683	40487-42-1	0.006	0.006	0.012	0.012	H	Dinitroaniline	–
Prometon	04037	1610-18-0	0.006	0.006	0.012	0.012	H	Triazine	–
Prometryn	04036	7287-19-6	0.003	0.003	0.006	0.006	H	Triazine	–
Propanil	82679	709-98-8	0.007	0.005	0.014	0.01	H	Chloroacetamide	–

Table A4. Reporting limits, uses, and classes of currently used pesticides and degradates analyzed in filtered wastewater-treatment-plant effluent or stormwater runoff, Columbia River Basin, Washington and Oregon, 2009–10.—Continued

[Shading indicates a detection in wastewater-treatment-plant effluent; bold type indicates a detection in stormwater runoff in this study. **CAS No.:** Chemical Abstract Service Registry Number® is a Registered Trademark of the American Chemical Society. CAS recommends the verification of the CASRNs through CAS Client Services℠. **Method detection limit** and **Reporting limit:** Values are in micrograms per liter. **Abbreviations:** NA, not available; D, degradate; Def, defoliant; F, fumigant; Fun, fungicide; H, herbicide; I, insecticide; N, nematocide; –, not applicable]

Analyte	Parameter code	CAS No.	Method detection limit		Reporting limit		Use	Class	Parent compound
			February–September 2009	October 2009–June 2010	February–September 2009	October 2009–June 2010			
Herbicides and degradates—Continued									
Propyzamide	82676	23950-58-5	0.002	0.0018	0.004	0.0036	H	Amide	–
Simazine	04035	122-34-9	0.005	0.003	0.01	0.006	H	Triazine	–
Tebuthiuron	82670	34014-18-1	0.01	0.014	0.02	0.028	H	Phenyl urea	–
Terbuthylazine	04022	5915-41-3	0.003	0.003	0.006	0.006	H	Triazine	–
Thiobencarb	82681	28249-77-6	0.008	0.008	0.016	0.016	H	Thiocarbamate	–
Trifluralin	82661	1582-09-8	0.006	0.009	0.012	0.018	H	Dinitroaniline	–
Insecticides and degradates									
Azinphos-methyl	82686	86-50-0	0.06	0.06	0.12	0.12	I	Organophosphate	–
Azinphos-methyl-oxon	61635	961-22-8	0.021	0.021	0.042	0.042	D	Organophosphate	Azinphos-methyl
Carbaryl	82680	63-25-2	0.1	0.03	0.2	0.06	I	Carbamate	–
Carbofuran	82674	1563-66-2	0.03	0.03	0.06	0.06	I	Carbamate	–
Chlorpyrifos	38933	2921-88-2	0.005	0.005	0.01	0.01	I	Organophosphate	–
Chlorpyrifos oxygen analog	61636	5598-15-2	0.025	0.025	0.05	0.05	D	Organophosphate	Chlorpyrifos
Cyfluthrin	61585	68359-37-5	0.008	0.008	0.016	0.016	I	Pyrethroid	–
lambda-Cyhalothrin	61595	91465-08-6	0.005	0.005	0.01	0.01	I	Pyrethroid	–
Cypermethrin	61586	52315-07-8	0.01	0.01	0.02	0.02	I	Pyrethroid	–
Desulfinylfipronil	62170	NA	0.006	0.006	0.012	0.012	D	Phenyl pyrazole	Fipronil
Desulfinylfipronil amide	62169	NA	0.015	0.015	0.029	0.029	D	Phenyl pyrazole	Fipronil
Diazinon	39572	333-41-5	0.003	0.003	0.005	0.005	I	Organophosphate	–
Diazinon oxygen analog	61638	962-58-3	0.0021	0.0021	0.006	0.006	D	Organophosphate	Diazinon
Dicrotophos	38454	141-66-2	0.04	0.04	0.08	0.08	I	Organothiophosphate	–
Dieldrin	39381	60-57-1	0.004	0.004	0.009	0.009	I	Organochlorine	–
Dimethoate	82662	60-51-5	0.003	0.003	0.006	0.006	I	Organothiophosphate	–
Disulfoton	82677	298-04-4	0.02	0.02	0.04	0.04	I	Organophosphate	–
Disulfoton sulfone	61640	2497-06-5	0.0068	0.0068	0.014	0.0136	D	Organophosphate	Disulfoton
alpha-Endosulfan	34362	959-98-8	0.003	0.003	0.006	0.006	I	Organochlorine	–
Endosulfan sulfate	61590	1031-07-8	0.011	0.007	0.022	0.014	D	Organochlorine	alpha-Endosulfan, beta-Endosulfan
Ethion	82346	563-12-2	0.006	0.004	0.012	0.008	I	Organothiophosphate	–
Ethion monoxon	61644	17356-42-2	0.011	0.011	0.021	0.021	D	Organothiophosphate	Ethion
Fipronil	62166	120068-37-3	0.02	0.009	0.04	0.018	I	Phenyl pyrazole	–
Fipronil sulfide	62167	120067-83-6	0.006	0.006	0.013	0.013	D	Phenyl pyrazole	Fipronil

Table A4. Reporting limits, uses, and classes of currently used pesticides and degradates analyzed in filtered wastewater-treatment-plant effluent or stormwater runoff, Columbia River Basin, Washington and Oregon, 2009–10.—Continued

[Shading indicates a detection in wastewater-treatment-plant effluent; bold type indicates a detection in stormwater runoff in this study. **CAS No.:** Chemical Abstract Service Registry Number® is a Registered Trademark of the American Chemical Society. CAS recommends the verification of the CASRNs through CAS Client Services^SM. **Method detection limit** and **Reporting limit:** Values are in micrograms per liter. **Abbreviations:** NA, not available; D, degradate; Def, defoliant; F, fumigant; Fun, fungicide; H, herbicide; I, insecticide; N, nematocide; –, not applicable]

Analyte	Parameter code	CAS No.	Method detection limit February–September 2009	Method detection limit October 2009–June 2010	Reporting limit February–September 2009	Reporting limit October 2009–June 2010	Use	Class	Parent compound
Insecticides and degradates—Continued									
Fipronil sulfone	62168	120068-36-2	0.012	0.012	0.024	0.024	D	Phenyl pyrazole	Fipronil
Fonofos	04095	944-22-9	0.005	0.0022	0.01	0.0044	I	Organophosphate	–
Isofenphos	61594	25311-71-1	0.003	0.003	0.006	0.006	I	Organothiophosphate	–
Malaoxon	61652	1634-78-2	0.04	0.04	0.08	0.08	D	Organophosphate	Malathion
Malathion	39532	121-75-5	0.01	0.008	0.02	0.016	I	Organophosphate	–
Methidathion	61598	950-37-8	0.003	0.003	0.006	0.006	I	Organothiophosphate	–
Methyl paraoxon	61664	950-35-6	0.005	0.005	0.01	0.01	D	Organophosphate	Methyl parathion
Methyl parathion	82667	298-00-0	0.004	0.004	0.008	0.008	I	Organophosphate	–
cis-Permethrin	82687	61949-76-6	0.007	0.007	0.014	0.014	I	Pyrethroid	–
Phorate	82664	298-02-2	0.01	0.01	0.02	0.02	I	Organophosphate	–
Phorate oxon	61666	2600-69-3	0.013	0.013	0.027	0.027	D	Organophosphate	Phorate
Phosmet	61601	732-11-6	0.1	0.017	0.2	0.034	I	Organothiophosphate	–
Phosmet oxon	61668	3735-33-9	0.0079	0.0079	0.0511	0.0511	D	Organothiophosphate	Phosmet
Propargite	82685	2312-35-8	0.01	0.01	0.02	0.02	I	Miscellaneous	–
Tefluthrin	61606	79538-32-2	0.005	0.005	0.01	0.01	I	Pyrethroid	–
Terbufos	82675	13071-79-9	0.009	0.009	0.018	0.018	I	Organophosphate	–
Terbufos oxygen analog sulfone	61674	56070-15-6	0.022	0.022	0.045	0.045	D	Organophosphate	Terbufos
Other compounds									
Dichlorvos	38775	62-73-7	0.01	0.01	0.02	0.02	F, D	Organophosphate	Naled
Ethoprop	82672	13194-48-4	0.008	0.008	0.016	0.016	N, I	Organophosphate	–
Fenamiphos	61591	22224-92-6	0.015	0.015	0.029	0.03	N	Organothiophosphate	–
Fenamiphos sulfone	61645	31972-44-8	0.027	0.027	0.053	0.053	D	Organothiophosphate	Fenamiphos
Fenamiphos sulfoxide	61646	31972-43-7	0.04	0.04	0.08	0.08	D	Organothiophosphate	Fenamiphos
1-Naphthol	49295	90-15-3	0.02	0.018	0.04	0.036	D	Carbamate, Chloroacetamide	Carbaryl, Napropamide
Tribufos	61610	78-48-8	0.018	0.009	0.035	0.018	Def	Organothiophosphate	–

Table A5. Reporting limits of polyaromatic hydrocarbons analyzed in unfiltered stormwater runoff, Columbia River Basin, Washington and Oregon, 2009–10.

[Bold type indicates a detection in stormwater runoff in this study. **CAS No.:** Chemical Abstract Service Registry Number® is a Registered Trademark of the American Chemical Society. CAS recommends the verification of the CASRNs through CAS Client Services[SM]. **Abbreviation:** μg/L, micrograms per liter]

Analyte	Parameter code	CAS No.	Reporting limit (μg/L) October 2008–September 2009	Reporting limit (μg/L) October 2009–September 2010	Analyte	Parameter code	CAS No.	Reporting limit (μg/L) October 2008–September 2009	Reporting limit (μg/L) October 2009–September 2010
Acenaphthene	34205	83-32-9	0.28	0.28	2,4-Dinitrophenol	34616	51-28-5	1.4	1.4
Acenaphthylene	34200	208-96-8	0.3	0.3	2,4-Dinitrotoluene	34611	121-14-2	0.6	0.56
Anthracene	34220	120-12-7	0.39	0.39	2,6-Dinitrotoluene	34626	606-20-2	0.43	0.4
Benzo[*a*]anthracene	34526	56-55-3	0.26	0.26	1,2-Diphenylhydrazine	82626	122-66-7	0.3	0.3
Benzo[a]pyrene	34247	50-32-8	0.33	0.33	**Di-*n*-butyl phthalate**	39110	84-74-2	1	2
Benzo[*b*]fluoranthene	34230	205-99-2	0.4	0.3	**Di-*n*-octyl phthalate**	34596	117-84-0	0.6	0.6
Benzo[*ghi*]perylene	34521	191-24-2	0.4	0.38	***bis*(2-Ethylhexyl) phthalate**	39100	117-81-7	2	2
Benzo[*k*]fluoranthene	34242	207-08-9	0.4	0.3	**Fluoranthene**	34376	206-44-0	0.3	0.3
4-Bromophenylphenylether	34636	101-55-3	0.36	0.24	**Fluorene**	34381	86-73-7	0.33	0.33
Butylbenzyl phthalate	34292	85-68-7	1.8	1.8	Hexachlorobenzene	39700	118-74-1	0.3	0.3
bis(2-Chloroethoxy)methane	34278	111-91-1	0.2	0.24	Hexachlorobutadiene	39702	87-68-3	0.2	0.24
bis(2-Chloroethyl)ether	34273	111-44-4	0.3	0.3	Hexachlorocyclopentadiene	34386	77-47-4	0.4	0.5
bis(2-Chloroisopropyl) ether	34283	108-60-1	0.38	0.14	Hexachloroethane	34396	67-72-1	0.2	0.24
4-Chloro-3-methylphenol	34452	59-50-7	0.55	0.55	**Indeno[*1,2,3-cd*]pyrene**	34403	193-39-5	0.4	0.38
2-Chloronaphthalene	34581	91-58-7	0.2	0.16	**Isophorone**	34408	78-59-1	0.4	0.26
2-Chlorophenol	34586	95-57-8	0.42	0.26	**Naphthalene**	34696	91-20-3	0.32	0.22
4-Chlorophenyl phenyl ether	34641	7005-72-3	0.34	0.34	Nitrobenzene	34447	98-95-3	0.2	0.26
Chrysene	34320	218-01-9	0.33	0.33	**2-Nitrophenol**	34591	88-75-5	0.4	0.4
Dibenz[*ah*]anthracene	34556	53-70-3	0.4	0.42	**4-Nitrophenol**	34646	100-02-7	0.51	0.51
1,2-Dichlorobenzene	34536	95-50-1	0.2	0.2	N-Nitrosodimethylamine	34438	62-75-9	0.2	0.24
1,3-Dichlorobenzene	34566	541-73-1	0.2	0.22	N-Nitrosodi-*n*-propylamine	34428	621-64-7	0.4	0.4
1,4-Dichlorobenzene	34571	106-46-7	0.2	0.22	**N-Nitrosodiphenylamine**	34433	86-30-6	0.4	0.28
3,3'-Dichlorobenzidine	34631	91-94-1	0.4	0.42	**Pentachlorophenol**	39032	87-86-5	1.2	0.6
2,4-Dichlorophenol	34601	120-83-2	0.39	0.36	**Phenanthrene**	34461	85-01-8	0.32	0.32
Diethyl phthalate	34336	84-66-2	0.61	0.61	**Phenol**	34694	108-95-2	0.44	0.28
Dimethyl phthalate	34341	131-11-3	0.4	0.36	**Pyrene**	34469	129-00-0	0.35	0.35
2,4-Dimethylphenol	34606	105-67-9	0.8	0.8	**1,2,4-Trichlorobenzene**	34551	120-82-1	0.2	0.26
4,6-Dinitro-2-methylphenol	34657	534-52-1	0.77	0.76	**2,4,6-Trichlorophenol**	34621	88-06-2	0.6	0.34

Table A6.   Reporting limits and methods used for trace elements analyzed in stormwater runoff, Columbia River Basin, Washington and Oregon, 2009–10.

[Bold type indicates a detection in stormwater runoff in this study. **CAS No.:** Chemical Abstract Service Registry Number® is a Registered Trademark of the American Chemical Society. CAS recommends the verification of the CASRNs through CAS Client Services[SM]. **Abbreviation:** μg/L, micrograms per liter]

Analyte	Parameter code	Method No.	CAS No.	Reporting limit (μg/L)	
				October 2008– September 2009	October 2009– September 2010
Unfiltered water					
Arsenic	01002	PLM11	7440-38-2	0.20	0.18
Cadmium	01027	PLM47	7440-43-9	0.06	0.04
Chromium	01034	PLM11	7440-47-3	0.40	0.42
Copper	01042	PLM11	7440-50-8	4.0	1.4
Lead	01051	PLM48	7439-92-1	0.10	0.06
Mercury	02708	CV018	7439-97-6	0.01	0.01
Nickel	01067	PLM11	7440-02-0	0.20	0.36
Selenium	01147	PLM11	7782-49-2	0.12	0.10
Silver	01077	PLM48	7440-22-4	0.06	0.016
Zinc	01092	PLA15	7440-66-6	4.0	4.0
Filtered water					
Arsenic	01000	PLM10	7440-38-2	0.06	0.044
Cadmium	01025	PLM43	7440-43-9	0.02	0.02
Chromium	01030	PLM10	7440-47-3	0.12	0.12
Copper	01040	PLM10	7440-50-8	1.0	1.0
Lead	01049	PLM43	7439-92-1	0.06	0.03
Mercury	02707	CV014	7439-97-6	0.01	0.01
Nickel	01065	PLM10	7440-02-0	0.12	0.12
Selenium	01145	PLM10	7782-49-2	0.06	0.04
Silver	01075	PLM43	7440-22-4	0.01	0.01
Zinc	01090	PLA11	7440-66-6	2.0	4.6

www.ingramcontent.com/pod-product-compliance
Lightning Source LLC
Chambersburg PA
CBHW081602170526
45166CB00009B/2793

* 9 7 8 1 5 0 0 4 9 1 2 2 2 *